SpringerBriefs in Applied Sciences and Technology

SpringerBriefs present concise summaries of cutting-edge research and practical applications across a wide spectrum of fields. Featuring compact volumes of 50 to 125 pages, the series covers a range of content from professional to academic.

Typical publications can be:

- A timely report of state-of-the art methods
- An introduction to or a manual for the application of mathematical or computer techniques
- A bridge between new research results, as published in journal articles
- A snapshot of a hot or emerging topic
- An in-depth case study
- A presentation of core concepts that students must understand in order to make independent contributions

SpringerBriefs are characterized by fast, global electronic dissemination, standard publishing contracts, standardized manuscript preparation and formatting guidelines, and expedited production schedules.

On the one hand, **SpringerBriefs in Applied Sciences and Technology** are devoted to the publication of fundamentals and applications within the different classical engineering disciplines as well as in interdisciplinary fields that recently emerged between these areas. On the other hand, as the boundary separating fundamental research and applied technology is more and more dissolving, this series is particularly open to trans-disciplinary topics between fundamental science and engineering.

Indexed by EI-Compendex, SCOPUS and Springerlink.

Hamidah Mohd Saman · Norhaliza Hamzah

Self Curing Concrete

Use of Green Artificial Aggregates (GAA) as Self Curing Agents

Hamidah Mohd Saman
School of Civil Engineering
College of Engineering
Universiti Teknologi MARA (UiTM)
Shah Alam, Selangor, Malaysia

Norhaliza Hamzah
UTM Construction Research Center
Universiti Teknologi Malaysia
Skudai, Johor, Malaysia

ISSN 2191-530X   ISSN 2191-5318   (electronic)
SpringerBriefs in Applied Sciences and Technology
ISBN 978-981-96-2963-3   ISBN 978-981-96-2964-0   (eBook)
https://doi.org/10.1007/978-981-96-2964-0

© The Editor(s) (if applicable) and The Author(s), under exclusive license to Springer Nature Singapore Pte Ltd. 2025

This work is subject to copyright. All rights are solely and exclusively licensed by the Publisher, whether the whole or part of the material is concerned, specifically the rights of translation, reprinting, reuse of illustrations, recitation, broadcasting, reproduction on microfilms or in any other physical way, and transmission or information storage and retrieval, electronic adaptation, computer software, or by similar or dissimilar methodology now known or hereafter developed.

The use of general descriptive names, registered names, trademarks, service marks, etc. in this publication does not imply, even in the absence of a specific statement, that such names are exempt from the relevant protective laws and regulations and therefore free for general use.

The publisher, the authors and the editors are safe to assume that the advice and information in this book are believed to be true and accurate at the date of publication. Neither the publisher nor the authors or the editors give a warranty, expressed or implied, with respect to the material contained herein or for any errors or omissions that may have been made. The publisher remains neutral with regard to jurisdictional claims in published maps and institutional affiliations.

This Springer imprint is published by the registered company Springer Nature Singapore Pte Ltd.
The registered company address is: 152 Beach Road, #21-01/04 Gateway East, Singapore 189721, Singapore

If disposing of this product, please recycle the paper.

Preface

As an academic in the field of concrete technology, I was motivated to write this book to share my knowledge on self cured concrete with researchers, professionals, students and practitioners. Self cured concrete offers numerous benefits, particularly in terms of environmental impact and sustainable development within the construction industry. By providing a comprehensive guide on this innovative material, I aim to contribute to the advancement of concrete technology and promote practices that support sustainability and environmental responsibility.

This book is structured into five comprehensive chapters that explore both conventional and self curing methods for concrete. It delves into the mechanisms, types of self curing agents and the properties of self cured concrete and mortar. Additionally, it examines the physical, deformation, microstructure and chemical properties of self curing agents, with a focus on the initial performance of concrete containing green artificial aggregates (GAAs) as self curing agents. Chapter 1 introduces the concept of self curing in concrete and its role in promoting sustainable development within the construction industry. Chapter 2 explains the fundamental principles of the cement hydration process and the mechanisms involved in the hydration of conventionally cured concrete. It also discusses the significance of self curing concrete, analyzing the environmental conditions and material properties that impact its efficiency and effectiveness.

Chapter 3 explores various types of self curing agents, including hydrophilic polymers, lightweight aggregates and chemical admixtures, each with unique mechanisms for facilitating internal curing. This chapter examines the effects of these agents on the physical, mechanical and durability properties of concrete and mortar, providing insights into optimizing performance through the integration of self curing agents. Chapter 4 investigates the physical properties, deformation characteristics, microstructure, interfacial transition zone (ITZ) and chemical properties of self cured concrete. It discusses key physical properties such as density, water absorption, water desorption, the shape and texture of the aggregates used. Lightweight aggregates (LWAs), for example, exhibit lower density due to their water-retaining properties. The chapter also reviews shrinkage as an essential parameter of self cured concrete.

Finally, Chap. 5 focuses on the physical and mechanical properties of green artificial aggregates (GAAs) as self curing agents. It compares the density, water absorption, desorption and texture of GAA with granite, providing valuable insights into the initial performance of concrete containing GAA versus conventional aggregates.

Shah Alam, Malaysia
Skudai, Malaysia

Hamidah Mohd Saman
Norhaliza Hamzah

Acknowledgements

First and foremost, I would like to express my sincere thanks to Associate Professor Dr. Rohana Hassan, Dr. Arni Munira Markom and Professor Dr. Aidah Jumahat for providing their invaluable guidance and encouragement in completing this book. Besides, I would like to extend my gratitude to my Mentor, Professor Dr. Azmi Ibrahim, for the insight guidance, advices imparting knowledge in the field of concrete structures, technology and material development.

I would like to acknowledge my Ph.D. student, Ms. Norhalizah Hamzah for the materials contributed in this book. This book will not be completed without her contribution. I also would like to express my gratitude to Universiti Teknologi MARA (UiTM) for providing facilities, laboratories and equipment for conducting research in which the review and findings are the major part of this book. Without these facilities, this book would not be possible.

Furthermore, I would like to thanks my colleagues for their love, motivation, encouragement and continuous support. Finally, I would like to thank my family members, my children Abdullah Mubarak, Fatimah, Safiah and Maimunah. Not forget, my husband Major Mohd Zakuan Mohd Nor for his love, caring, inspiration and continuous support who motivates me when in need. This blessing has enlightened my path toward completing this book.

Lastly, I am very thankful to all researchers whose research papers, thesis, book and articles that have been referred, without these references. I may not able to get right path for my book.

About This Book

This book delves into various methods of self curing concrete, offering an in-depth review that serves as a valuable reference for both students and researchers in the field. Its unique approach lies in the use of simple, clear English, making it accessible to readers of all levels. By breaking down complex concepts into understandable terms, the book ensures that even those new to the topic can grasp the fundamentals. Additionally, it is highly relevant for practitioners, providing practical insights into different types of self curing techniques that can be implemented in real-world applications. Whether you are an academic looking to expand your knowledge or a professional seeking practical solutions, this book offers guidance on self curing concrete, making it an essential resource for anyone interested in this innovative area of construction technology. The authors hope this book will give valuable insights into the development of lightweight artificial aggregates and other potential materials as self curing agents.

Contents

1 **Curing of Concrete and Its Main Issues** 1
  1.1 Introduction 1
  1.2 Importance of Curing 1
  1.3 Conventional Curing Methods and Issues 2
  1.4 Emergence of Self Curing Concrete 4
  References 6

2 **Mechanisms of Conventional and Self Cured Cement Hydration** 11
  2.1 Introduction 11
  2.2 Cement Hydration of Conventional Concrete 11
  2.3 Stages of Hydration Process 14
    2.3.1 Initial Mixing and Pre-induction Period 14
    2.3.2 Induction Period 14
    2.3.3 Acceleration Period 14
    2.3.4 Deceleration and Steady-State Periods 15
  2.4 Microstructural Development of Hydration Products 15
    2.4.1 Formation of C–S–H Gel 16
    2.4.2 Growth of C–H Crystals 16
    2.4.3 Formation of Ettringite and Monosulfoaluminate 16
    2.4.4 Pore Structure and Capillary Porosity 17
  2.5 Factors Influencing Hydration 17
    2.5.1 Water-Cement Ratio (W/C) 17
    2.5.2 Temperature 17
    2.5.3 Cement Composition 18
    2.5.4 Admixtures 18
  2.6 Mechanisms of Self Curing Concrete 18
  2.7 Significance of Self Curing Concrete 20
  2.8 Future Studies in Self Curing Concrete 22
  References 23

3 Types of Self Curing Agents and Properties of Self Cured Concrete/Mortar ... 25
   3.1 Introduction ... 25
   3.2 Types of Self Curing Agents and Mechanism in Cementitious Materials ... 25
      3.2.1 Artificial Lightweight Aggregate ... 26
      3.2.2 Porous Superfine Powders ... 27
      3.2.3 Superabsorbent Polymer (SAP) ... 28
      3.2.4 Polyethylene Glycol (PEG) ... 30
      3.2.5 Natural Fibers ... 30
      3.2.6 Artificial Normal Weight Aggregate (ANWA) ... 31
   3.3 Effects of Self Curing Agents on Properties of Concrete or Mortar ... 33
      3.3.1 Workability ... 33
      3.3.2 Density ... 34
      3.3.3 Compressive Strength ... 34
      3.3.4 Split Tensile and Flexural Strength ... 40
   References ... 43

4 Physical, Deformation, Microstructure and Chemical Properties of Self Cured Agents and Concrete ... 53
   4.1 Introduction ... 53
   4.2 Physical Properties of Lightweight Aggregate (LWA) as Self Curing Agents ... 53
      4.2.1 Water Absorption ... 54
      4.2.2 Water Desorption ... 54
      4.2.3 Shape and Surface Texture ... 55
   4.3 Autogeneous Shrinkage and Drying Shrinkage of Self Cured Concrete ... 55
   4.4 Microstructure Examination Using Scanning Electron Microscope (SEM) ... 59
   4.5 Effect of Curing Agent on Interfacial Transition Zone (ITZ) ... 60
   4.6 Chemical Analysis Using X-Ray Diffraction (XRD) ... 62
   References ... 66

5 Characteristics of Green Artificial Aggregates (GAA) as Self Curing Agent ... 71
   5.1 Introduction ... 71
   5.2 Physical Properties of GAA ... 72
      5.2.1 Specific Gravity ... 72
      5.2.2 Water Absorption ... 73
   5.3 Water Desorption ... 73
   5.4 Morphology and Microstructure of Aggregate ... 75
   5.5 Mercury Intrusion Porosimetry (MIP) ... 76
   5.6 Aggregate Crushing Value (ACV) ... 76

5.7	Initial Performance of GAA as Self Cured Agent in Concrete		77
	5.7.1	Workability of Concrete with and Without GAA	78
	5.7.2	Compressive Strength of Concrete with and Without GAA	79
References			82

Index 85

List of Figures

Fig. 2.1	Electron microscopic images of hardened cement paste after hydration (Stutzman 2001; used with permission of John Wiley and Sons, from Materials Science of Concrete, Special Volume: Calcium Hydroxide in Concrete, Copyright 2001; permission conveyed through Copyright Clearance Center, Inc.)	13
Fig. 2.2	Illustration of the differences between self curing and external curing (Bentz and Weiss 2011)	19
Fig. 2.3	Mechanism of self curing concrete (Sampebulu' 2012; licensed under a Creative Commons Attribution (CC BY))	20
Fig. 2.4	Contact zone under internal curing and normal curing (Han et al. 2017; Reproduced with permission from Springer Nature)	21
Fig. 3.1	Model moving of water in concrete incorporate self curing with $r(t) <$ Ra (Nguyen and Le 2018; licensed under a Creative Commons Attribution (CC BY))	26
Fig. 3.2	Microstructure of cenopheres under SEM observation (Liu et al. 2019; licensed under a Creative Commons Attribution (CC BY))	28
Fig. 3.3	Process of water uptake to SAP (Mechtcherine and Reinhardt 2012; licensed under a Creative Commons Attribution (CC BY))	29
Fig. 3.4	Hydrogen bonds between water molecules and an –OH group on a polymer molecule (Dhir et al. 1994; Reproduced with permission from Springer Nature)	31
Fig. 3.5	Free and bound water in wood (adapted from Ahmed 2006)	32
Fig. 3.6	Compressive strength of porous aggregate with percent replacement of conventional aggregate at 28 days	35

Fig. 3.7	Compressive strength for concrete containing different content of lightweight aggregate at 7 and 28 days (Reprinted from Gopi and Revathi 2021, Copyright 2021, with permission from Elsevier)	37
Fig. 3.8	a Soaked SAPs in the cement mixture; b SAPs gradually release the absorbed water; c SAPs de-swell and leave voids (Reprinted (adapted) with permission from Erk and Bose 2018. Copyright 2018 American Chemical Society)	39
Fig. 3.9	Compressive strength of UHPC with varying proportions of LWAs: a at 7, 28, and 56 days of curing (Reproduced from Abadel 2023; licensed under a Creative Commons Attribution (CC BY))	40
Fig. 3.10	Flexural strength of UHPC containing lightweight aggregate at 7, 28, and 56 days of curing (Abadel 2023; licensed under a Creative Commons Attribution (CC BY))	42
Fig. 4.1	Effect of autogenous shrinkage of concrete with air-dry (AD) and saturated-surface-dry (SSD) lightweight coarse aggregate (Bentur et al. 2001, Copyright 2001, with permission from Elsevier)	58
Fig. 4.2	Effect of LWA on shrinkage of UHPC specimens (Abadel 2023)	59
Fig. 4.3	Normal-bonded (left) and well-bonded AAC-LWA (right) with cement paste on interfacial transition zone (ITZ) (Suwan and Wattanachai 2017)	61
Fig. 4.4	ITZ microstructure in cement mortar a with porous aggregate, b with conventional aggregate (Reprinted from Sun et al. 2015, Copyright 2015, with permission from Elsevier)	61
Fig. 4.5	The microstructure of the high-performance concrete with SAP (Reprinted from Liu et al. 2020, Copyright 2020, with permission from Elsevier)	62
Fig. 4.6	X-ray diffraction peaks of SCC subjected to different curing regimes at 27 ± 2 °C (used with permission of Madduru et al. 2018; permission conveyed through Copyright Clearance Center, Inc.)	63
Fig. 4.7	X-ray diffraction analysis (Tang 2017)	64
Fig. 4.8	XRD spectra of UHPC: a LW0, b LW15, c LW30 (Abadel 2023)	65
Fig. 5.1	Water absorption of GAA after 24 h soaked in pigmented water	73
Fig. 5.2	The water movement in the concrete specimen from pre-wetted GAA to cement paste	74
Fig. 5.3	Water desorption properties of pre-wetted GAA at different RHs	75
Fig. 5.4	The model illustrating the movement of water from pre-wetted GAA to cement paste	75

Fig. 5.5	Image of **a** GAA and **b** granite up to 4.5 × magnification	76
Fig. 5.6	Pore size distribution of GAA	77
Fig. 5.7	Pore size distribution of granite (Gao et al. 2021)	77
Fig. 5.8	Slump (in mm) of fresh concrete mixes containing 50 and 100% GAA which prior to that the aggregates were made SSD and AD conditions	79
Fig. 5.9	Compressive strength of the concrete specimens containing SSD, AD GAA cured in air and water curing	80

List of Tables

Table 3.1	Compressive strength reduction of concrete due to LWA addition as compared to control for 28 days (Akhnoukh 2017)	36
Table 3.2	Compressive strength increment and decrement of concrete due to LWA addition as compared to control for 28 days (Suwan and Wattanachai 2017)	37
Table 3.3	Compressive strength of mixtures with SAP at 28 days	39
Table 3.4	Flexural strength and split tensile strength for concrete specimen at 28 days (Pradeep et al. 2019)	41
Table 3.5	Effect of LECA on different mechanical properties of concrete at 28 days (Mousa et al. 2015b)	41
Table 5.1	Physical, mechanical properties standard used for testings coarse aggregates	72
Table 5.2	The percentage of increased concrete strength in 3 days, 7 days and 28 days cured in water curing (WC)	81
Table 5.3	The increase (%) in concrete strength in 3 days, 7 days and 28 days cured in air curing (AC)	81

Chapter 1
Curing of Concrete and Its Main Issues

1.1 Introduction

In the construction industry, curing concrete is a critical process that significantly influences the final strength, durability and overall performance of the material. This chapter delves into the essential role of curing in ensuring that cast concrete achieves its desired properties. It highlights the conventional curing methods commonly used, such as water curing, membrane curing and steam curing, while discussing the main issues and limitations associated with these techniques.

The chapter then introduces the concept of self curing, an innovative and alternative approach designed to enhance the performance and longevity of concrete. Self curing, also known as internal curing, addresses the shortcomings of traditional methods by providing a continuous supply of moisture from within the concrete itself. Furthermore, the chapter explores the role of self curing in promoting sustainable development within the construction industry. By reducing the need for external water sources and minimizing water wastage, self curing techniques offer significant environmental benefits. The chapter provides an overview of how adopting self curing methods can lead to more eco-friendly construction practices, aligning with the industry's goals of sustainability and environmental stewardship.

Through the explanation, this chapter aims to provide an understanding of curing methods and their impact on concrete, emphasizing the advantages of self curing as a forward-thinking solution for modern construction challenges.

1.2 Importance of Curing

Concrete is the most used construction material worldwide, comprised of Portland cement, water and aggregates. Fine aggregates, water and cement blended together to form a paste that coats all the individual pieces of coarse aggregate, thereby, resulting in a plastic mixture. Concrete, a ubiquitous building material, owes its strength and durability to the hydration of cementitious materials in the presence of water. A chemical reaction called hydration occurs between water and cement. This fundamental chemical process forms a hardened matrix that supports structures ranging from residential homes to towering infrastructure. Curing allows continuous hydration of cement until achieving its potential strength and durability. A concrete turns to a solid state from a plastic state for approximately 2 h (with chemical admixture), and it continues to obtain its strength as it cures. Curing is the process of controlling the rate, extent of moisture loss and temperature in concrete during cement hydration and reduces water evaporation (James et al. 2011; Neville 2011; Mohamad et al. 2017; Nahata et al. 2014; Vázquez-Rodríguez et al. 2020; Yang et al. 2022). It is crucial to ensure the condition moisture is proper, otherwise, the hydration of cement virtually ceases due to relative humidity (RH) within the capillaries drops below 80% (Parrott and Illston 1975; Spears 1983; Neville 2011). If hydration stopped, sufficient calcium silicate hydrate (C–S–H) cannot be developed (Rajappan et al. 2014; Shahbazpanahi et al. 2021; Tang et al. 2021), which disrupts the development of dense microstructure and the refined pore structure within the cement matrices allowing the ingress of deleterious agents into the concrete. These, subsequently lead to poor quality of concrete, such as causing plastic shrinkage cracks, poorly formed hydrated products, finishing issues and other surface defects (Mohamad et al. 2017; McCarter and Ben-Saleh 2001; Ye et al. 2010).

In addition, if concrete is not properly cured, then the surface layer, about 30–50 mm (Gowripalan et al. 1990; Neville 2011; Taylor 2014) is most affected due to the potential for excessive evaporation of water from the concrete surface. Thus, the moisture loss leads to lower degree of cement hydration and a porous zone between the aggregates and cement paste, resulting in undesirable concrete properties such as decreased compressive strength and increased porosity (Sathanandham et al. 2013; Nduka et al. 2018; Bandara et al. 2019). In sum, curing concrete is a crucial practice in construction to gain optimal performance in any environment or application.

1.3 Conventional Curing Methods and Issues

Conventionally, achieving optimal concrete properties necessitates meticulous curing practices to maintain adequate moisture levels during the critical early stages of hydration. Curing is a process of maintaining the rate and the extent of moisture loss within a proper temperature in concrete during cement hydration and reduces water evaporation (Huseien et al. 2021; James et al. 2011; Nahata et al. 2014; Mohamad et al. 2017). There are several methods of conventional curing that are commonly used in the construction industry. Conventional curing methods, predominantly, water curing and the application of curing compounds, have long been employed to mitigate moisture loss and promote the development of concrete's mechanical properties.

It has typically been common to remove formwork within 24 or 48 h of concreting on-site, which is subsequently aided by conventional curing methods to further

1.3 Conventional Curing Methods and Issues

develop the desired properties of concrete. Throughout the building, concrete is cast in a variety of shapes and locations. Continuous curing is challenging for concrete applied at heights, vertical members, pavement, sloped roofs and concrete with a wide thickness (Chand et al. 2015; Nduka et al. 2018; Rodríguez-Torres and Torres-Castellanos 2019).

One of the most common methods is wet curing, which involves keeping the surface of the concrete moist by applying a continuous moist of water or covering it with wet burlap or other similar materials (Rodríguez-Torres and Torres-Castellanos 2019; Bandara et al. 2019; Gereziher and Zhutovsky 2022). Water curing involves keeping the concrete surface continuously moist through ponding, sprinkling, or wet coverings, which slows down moist evaporation and minimizes cracking due to shrinkage. Curing compounds, on the other hand, form a membrane over the concrete surface, reducing water evaporation and retaining moisture within the concrete. Another method is steam curing, which involves applying heat and moisture to the concrete using steam (Halit et al. 2005; Zou et al. 2018; Mei et al. 2018; Uygunoğlu and Hocaoğlu 2018).

However, water curing demands substantial quantities of water, which may not be readily available in arid regions or may contribute to logistical complexities and increased construction costs. Curing compounds, while offering convenience, must be applied correctly to ensure uniform coverage and effectiveness, adding to project timelines and costs. Furthermore, both methods require careful monitoring and management to prevent premature drying and ensure consistent concrete quality.

Previous studies have reported various methods to prevent the loss of moisture during concreting like spraying or fogging the surface of new cast concrete with water (ACI Commitee 308 2001; Bushlaibi 2004; Leger et al. 2004; Neville 2011; Wasserman and Bentur 2013; Akinwumi and Gbadamosi 2014), using wet surface covering (Neville 2011; ACI Commitee 308 2001; Mccarter and Ben-Saleh 2001; Bushlaibi 2004; Wasserman and Bentur 2013; Nahata et al. 2014; Akinwumi and Gbadamosi 2014) and water ponding which is suitable for horizontal surfaces (Neville 2011; ACI Commitee 308 2001; Leger et al. 2004; Wasserman and Bentur 2013; Akinwumi and Gbadamosi 2014; Nahata et al. 2014). Other methods include membrane curing which keeps the water within the concrete to maximize the potential hydration (Ye et al. 2010; Wasserman and Bentur 2013; Akinwumi and Gbadamosi 2014; Nahata et al. 2014), steam (ACI Commitee 308 2001; Halit et al. 2005; Gonzalez-Corominas et al. 2016; Mei et al. 2018) and leaving formwork in place (Neville 2011; Leger et al. 2004). The accelerated curing method such as microwave curing (Prommas and Rungsakthaweekul 2014; Rattanadecho et al. 2016; Mgbemena et al. 2018; Hamada et al. 2022), direct electric curing (Heritage 2001; Kovtun et al. 2016; Cecini et al. 2018). Infrared curing (Gambhir 2013) is used in application of heat on fresh concrete to promote the rapid cement hydration by gaining the early-age strength of concrete. However, a common feature of all the existing curing techniques is often applied to the surface external of concrete. External curing addresses only the proper hydration on surface of concrete. The internal part of the concrete, on the other hand, is not completely hydrated. If the capillary porosity in the concrete disconnected during the curing process, moisture would be unable to penetrate the full depth of the concrete and causing limiting on effectiveness in curing concrete.

In sum, despite the effectiveness of the conventional curing methods, these conventional methods, however, come with practical challenges and resource implications, prompting the exploration of alternative approaches such as self curing concrete which is discussed in the following section.

1.4 Emergence of Self Curing Concrete

Self curing concrete or internal curing can be defined as concrete with additional water capacity during the curing regime for the hydration process (El-Dieb 2007). The practice of self curing is a probable technique that can supply more water to concrete for more effective cement hydration and decreased self-desiccation. Therefore, several researchers introduced the concept of self curing concrete (Dhir et al. 1994a, b, 1995, 1996; Reinhardt and Weber 1998). The concept of self curing concrete is introduced to mitigate the process of water evaporation in concrete and hence, increase the capacity of water retention in concrete compared to conventional curing concrete (Reinhardt and Weber 1998; Bentz et al. 2005; El-Dieb 2007; Zhutovsky et al. 2011; Thrinath and Sundara 2017; Vázquez-Rodríguez et al. 2020).

The subsequent chapters discuss the types of curing agents and previous studies related to the influence of curing agents in the self curing cement-based system. Generally, all types of self curing agents have shown promising results in the self curing cement-based systems and have the potential to extend to new studies.

In response to the challenges of conventional curing highlighted in Sect. 1.3, self curing concrete has emerged as a promising alternative (Dhir et al. 1994a, b, 1998). The concept of self curing involves integrating specific materials or additives directly into the concrete mix to facilitate internal moisture retention during hydration. This innovative approach aims to reduce or eliminate the need for external curing processes, thereby, enhancing construction efficiency, sustainability and overall concrete performance. Therefore, the technique in which water-filled internal curing agents reacts as reservoirs are added to concrete mixture which can gradually release its water during hydration and evaporation process (Jongvisuttisun et al. 2018; Sastry and Putturu 2018; Mrad and Chehab 2019; Meyst et al. 2020) which is the main mechanistic of self curing concrete and will be further explained in the following paragraphs and subsequent chapters.

Generally, lightweight aggregate (LWA) used in cement-based materials is able to replace normal granite or sand for structural application (Aslam et al. 2016; Serri et al. 2016; Etxeberria 2020). Previous researchers were interested in investigating the physical and mechanical properties of high-performance concrete (HPC) by replacing natural aggregates with the partial replacement of LWA as a curing agent (Liu et al. 2017; Zhang and Poon 2017; Zhutovsky and Konstantin 2017; Kim and Lee 2018; Zou et al. 2018; Chen et al. 2019). However, limited studies have been carried out on replacing natural aggregates with 100% artificial aggregate produced from the

1.4 Emergence of Self Curing Concrete

combination of wastes as curing agents. For that reason, there is a need to assess the self cured concrete using green artificial aggregate (GAA) as a curing agent under external curing and without external curing process in the normal strength concrete for comparison.

Self curing agents operate through various mechanisms tailored to address specific hydration dynamics and environmental conditions. Superabsorbent polymers (SAPs), for instance, possess the ability to absorb large amounts of water and release it gradually within concrete matrix, effectively maintaining moisture levels critical for hydration. Internal curing aggregates, such as LWAs pre-saturated with water, serve as reservoirs that supply moisture as needed during cement hydration. These internal mechanisms not only optimize hydration kinetics but also contribute to mitigating shrinkage and cracking, enhancing the long-term durability and structural integrity of concrete.

Many researchers have studied self curing concrete composing using different materials that have potential as self curing agents which are LWA such as expanded shale, clay and slate, pumice and bottom ash (Zhutovsky and Konstantin 2017; Akhnoukh 2017; Kevern and Nowasell 2018; Zou et al. 2018; Kim and Lee 2018; Vázquez-Rodríguez et al. 2020), superabsorbent polymers (Riyazi et al. 2017; Almeida and Klemm 2018; Kang et al. 2018; Oh and Young 2018; Woyciechowski and Maciej 2018; Liu et al. 2021), polyethylene glycol (Bashandy 2016; Madduru et al. 2016, 2018; Sarbapalli et al. 2017; Amin et al. 2021), natural fibers (Jongvisuttisun et al. 2018; Zadeh and Bobko 2014; Kawashima and Surendra 2011; Dávila-Pompermayer et al. 2020) and artificial normal weight aggregate such as ceramic and recycle concrete waste (Aslam et al. 2016; Gonzalez-Corominas and Etxeberria 2016; Shigeta et al. 2018; Etxeberria 2020). The mechanisms and properties of these self curing agents will be explained in Chap. 3.

Artificial aggregates produced from several waste materials could be one of agents that could be used as self cured agent in concrete. Green artificial aggregate (GAA) is a production from a combination of the wastes, for example, ceramic waste, bottom ash and fly ash. The influence of incorporating GAA as self curing agent to concrete properties and the mechanism under curing and without proper curing process is becoming the interest of this book. Within the knowledge of the authors, the application of artificial aggregate produced from combination of wastes to be used as self curing agent in concrete has not been reported elsewhere. Therefore, it is an intention of the authors to explore the utilization of GAA, one of LWA as self curing agent in concrete. It drives the authors to investigate the influence of incorporating GAA as self curing agent toward the densification of internal microstructure for normal strength concrete by using Scanning Electron Microscopy (SEM), X-ray diffraction (XRD) and Mercury Intrusion Porosimetry (MIP) which will be elaborated in Chaps. 4 and 5.

In conclusion, this introductory chapter sets the stage for an in-depth exploration of self curing concrete, highlighting its evolution as a viable solution to challenges in implementing conventional curing. By examining the principles, mechanisms and applications of self curing agents, this book aims to equip researchers, practitioners and stakeholders with the knowledge and insights needed to advance the field of concrete technology toward more sustainable and resilient built environments.

This book also provides a foundational overview of self curing concrete, emphasizing its context within conventional curing methods and emergence as an innovative solution. It lays the groundwork to explore specific aspects of self curing agents in greater detail.

References

ACI Commitee 308, *Guide to Curing Concrete* (American Concrete Institute, 2001)
I.I. Akinwumi, Z. Gbadamosi, Effects of curing condition and curing period on the compressive strength development of plain concrete. Int. J. Civil Environ. Res. **1**(2), 83–99 (2014)
A.K. Akhnoukh, Internal curing of concrete using lightweight aggregates. Part. Sci. Technol. **36**(3), 362–367 (2017). https://doi.org/10.1016/j.conbuildmat.2017.11.055
C.R. Almeida Fernando, J.K. Agnieszka, Efficiency of internal curing by Superabsorbent Polymers (SAP) in PC-GGBS mortars. Cem. Concr. Compos. **88**, 41–51 (2018). https://doi.org/10.1016/j.cemconcomp.2018.01.002
M. Amin, A.M. Zeyad, B.A. Tayeh, I.S. Agwa, Engineering properties of self-cured normal and high strength concrete produced using polyethylene glycol and porous ceramic waste as coarse aggregate. Constr. Build. Mater. **299**(June), 124243 (2021). https://doi.org/10.1016/j.conbuildmat.2021.124243
M. Aslam, P. Shafigh, Z.J. Mohd, L. Mohamed, Benefits of using blended waste coarse lightweight aggregates in structural lightweight aggregate concrete. J. Clean. Prod. **119**, 108–117 (2016). https://doi.org/10.1016/j.jclepro.2016.01.071
M.H.W. Bandara, W.K. Mampearachchi, T. Anojan, Enhance the properties of concrete using pre-developed burnt clay chips as internally curing concrete aggregate. Case Stud. Constr. Mater.**11**. https://doi.org/10.1016/j.cscm.2019.e00284
A.A. Bashandy, Self-Curing concrete under sulfate attack. Arch. Civil Eng. LXII **2**, 3–18 (2016)
D.P. Bentz, P. Lura, J.W. Roberts, Mixture proportioning for internal curing. Concr. Int. **27**(2), 35–40 (2005)
A.H.Ã. Bushlaibi, Effects of environment and curing methods on the compressive strength of silica fume high-strength concrete. Adv. Cem. Res. **16**(1), 17–22 (2004)
D. Cecini, S. Austin, S. Cavalaro, A. Palmeri, Accelerated electric curing of steel-fibre reinforced concrete. Constr. Build. Mater. **189**, 192–204 (2018). https://doi.org/10.1016/j.conbuildmat.2018.08.183
M.S.R. Chand, S.N.R.G. Pollapothu, R.K. Garje, R.K. Pancharathi, Paraffin wax as an internal curing agent in ordinary concrete. Mag. Concr. Res. **67**(2), 82–88 (2015). https://doi.org/10.1680/macr.14.00192
F. Chen, W. Kai, R. Lijian, X. Jianan, Z. Huiming, Internal curing effect and compressive strength calculation of recycled clay brick aggregate concrete. Materials **12**(11), 14 (2019)
R.L.G. Dávila-Pompermayer, P. Lopez-Yepez, C.A.J. Valdez-Tamez, A. Durán-Herrera, Lechugilla natural fiber as internal curing agent in self compacting concrete (SCC): mechanical properties, shrinkage and durability. Cem. Concr. Compos. **112**, 103686 (2020). https://doi.org/10.1016/j.cemconcomp.2020.103686
R.K. Dhir, C. Peter, P.C. Hewlett, J.S. Lota, D.D. Thomas, An investigation into the feasibility of formulating 'self-cure' concrete. Mater. Struct. **27**, 606–615 (1994a)
R.K. Dhir, P.C. Hewlett, J.S. Lota, T.D. Dyer, An investigation into the feasibility of formulating 'self-cure' concrete. Mater. Struct. **27**, 606–615 (1994b)

References

R.K. Dhir, P.C. Hewlett, T.D. Dyer, Durability of 'self-cure' concrete. Cem. Concr. Res. **25**(6), 1153–1158 (1995)

R.K. Dhir, P.C. Hewlett, T.D. Dyer, Influence of microstructure on the physical properties of self-curing concrete. ACI Mater. J. **93**(5), 465–471 (1996)

R.K. Dhir, P.C. Hewlett, T.D. Dyer, Mechanisms of water retention in cement pastes containing a self-curing agent. Mag. Concr. Res. **50**, 85–90 (1998)

A.S. El-Dieb, Self-curing concrete: water retention, hydration and moisture transport. Constr. Build. Mater. **21**, 1282–1287 (2007). https://doi.org/10.1016/j.conbuildmat.2006.02.007

M. Etxeberria, in *Advances in Construction and Demolition Waste Recycling*. The Suitability of Concrete Using Recycled Aggregates (RAs) for High-Performance Concrete (Elsevier Ltd., 2020). https://doi.org/10.1016/b978-0-12-819055-5.00013-9.

M.L. Gambhir, *Concrete Technology: Theory and Practice* (McGraw Hill Education (India) Private Limited, 2013)

A.T. Gereziher, S. Zhutovsky, The effect of external curing methods on the development of mechanical and durability-related properties of normal-strength concrete. Constr. Build. Mater. **324**(February), 126706 (2022). https://doi.org/10.1016/j.conbuildmat.2022.126706

A. Gonzalez-Corominas, M. Etxeberria, Effects of using recycled concrete aggregates on the shrinkage of High Performance Concrete. Constr. Build. Mater. **115**, 32–41 (2016). https://doi.org/10.1016/j.conbuildmat.2016.04.031

N. Gowripalan, J.G. Cabrera, A.R. Cusens, P.J. Wainwright, Effect of curing on durability. Concr. Int. 47–54 (1990)

H. Hamada, A. Alattar, B. Tayeh, F. Yahaya, I. Almeshal, Case studies in construction materials influence of different curing methods on the compressive strength of ultra-high-performance concrete: a comprehensive review. Case Stud. Constr. Mater. **17**(August), e01390 (2022). https://doi.org/10.1016/j.cscm.2022.e01390

Y. Halit, A. Serdar, Y. Huseyin, B. Baradan, Effect of steam curing on Class C high-volume fly ash concrete mixtures. Cem. Concr. Res. **35**, 1122–1127 (2005). https://doi.org/10.1016/j.cemconres.2004.08.011

I. Heritage, *Direct Electric Curing of Mortar and Concrete*. A thesis submitted in partial fulfilment of the requirements of Napier University, for the degree of Doctor of Philosophy. School of the Built Environment, Napier University, Edinburgh, UK (2001)

G.F. Huseien, A.R.M. Sam, H.A. Algaifi, R. Alyousef, Development of a sustainable concrete incorporated with effective microorganism and fly ash: Characteristics and modeling studies. Constr. Build. Mater. **285**, 122899 (2021)

T.A. James, E.W. Malachi, Gadzama, V. Anametemfiok, Effect of curing methods on the compressive strength of concrete. Nigerian J. Technol. **30**(3), 14–20 (2011). https://doi.org/10.18535/ijecs/v5i7.09

P. Jongvisuttisun, L. Johannes, E.K. Kimberly, Key mechanisms controlling internal curing performance of natural fibers. Cem. Concr. Res. **107**, 206–220 (2018). https://doi.org/10.1016/j.cemconres.2018.02.007

S.H. Kang, H. Sung-gul, M. Juhyuk, Importance of drying to control internal curing effects on field casting ultra-high performance concrete. Cem. Concr. Res. **108**, 20–30 (2018). https://doi.org/10.1016/j.cemconres.2018.03.008

S. Kawashima, P.S. Surendra, Early-age autogenous and drying shrinkage behavior of cellulose fiber-reinforced cementitious materials. Cem. Concr. Compos. **33**(2), 201–208 (2011). https://doi.org/10.1016/j.cemconcomp.2010.10.018

J.T. Kevern, Q.C. Nowasell, Internal curing of pervious concrete using lightweight aggregates. Constr. Build. Mater. **161**, 229–235 (2018). https://doi.org/10.1016/j.conbuildmat.2017.11.055

H.K. Kim, H.K. Lee, Hydration kinetics of high-strength concrete with untreated coal bottom ash for internal curing. Cem. Concr. Compos. **91**, 67–75 (2018). https://doi.org/10.1016/j.cemconcomp.2018.04.017

M. Kovtun, M. Ziolkowski, J. Shekhovtsova, E. Kearsley, Direct electric curing of alkali-activated fly ash concretes: a tool for wider utilization of fly ashes. J. Clean. Prod. **133**, 220–227 (2016). https://doi.org/10.1016/j.jclepro.2016.05.098

E. Leger, M. Miller, R. Miller, *Complete Building Construction* (Wiley Publishing, Inc, 2004). http://www.getcited.org/pub/102977715

J. Liu, S. Caijun, M. Xianwei, H.K. Kamal, Z. Jian, W. Dehui, An overview on the effect of internal curing on shrinkage of high performance cement-based materials. Constr. Build. Mater. **146**, 702–712 (2017). https://doi.org/10.1016/j.conbuildmat.2017.04.154

J. Liu, F. Nima, H.K. Kamal, S. Caijun, Effects of SAP characteristics on internal curing of UHPC matrix. Constr. Build. Mater. **280** (2021). https://doi.org/10.1016/j.conbuildmat.2021.122530

S.R.C. Madduru, N.R.G.P. Swamy, K.P. Rathish, K.G. Rajesh, C. Raveena, Effect of Self curing chemicals in self compacting mortars. Constr. Build. Mater. **107**, 356–364 (2016). https://doi.org/10.1016/j.conbuildmat.2016.01.018

S.R.C. Madduru, K.P. Rathish, N.R.G.P. Swamy, K.G. Rajesh, Performance and microstructure characteristics of self-curing self-compacting concrete. Adv. Cem. Res. **30**(10), 451–468 (2018). https://doi.org/10.1680/jadcr.17.00154

W.J. McCarter, A.M. Ben-Saleh, Influence of practical curing methods on evaporation of water from freshly placed concrete in hot climates. Build. Environ. **36**, 919–924 (2001)

C.O. Mgbemena, D. Li, M. Lin, P. Daniel, K.B. Katnam, V.T. Kumar, H.Y. Nezhad, Accelerated microwave curing of fibre-reinforced thermoset polymer composites for structural applications: a review of scientific challenges. Compos. A Appl. Sci. Manuf. **115**, 88–103 (2018). https://doi.org/10.1016/j.compositesa.2018.09.012

J. Mei, M. Baoguo, T. Hongbo, L. Hainan, L. Xiaohai, J. Wenbin, Z. Ting, Influence of steam curing and nanosilica on hydration and microstructure characteristics of high volume fly ash cement system. Constr. Build. Mater. **171**, 83–95 (2018). https://doi.org/10.1016/j.conbuildmat.2018.03.056

L.D. Meyst, K. Judy, R.T.F. José, V.T. Kim, D.B. Nele, The use of superabsorbent polymers in high performance concrete to mitigate autogenous shrinkage in a large-scale demonstrator. Sustainability **12**(11) (2020). https://doi.org/10.3390/su12114741

D. Mohamad, S. Beddu, S.N. Sadon, N.L.M. Kamal, Z. Itam, M.A. Zainol, M.Z. Ramli, W.M. Sapuan, Properties of self curing concrete containing bottom ash. Int. J. Adv. Appl. Sci. **4**, 138–142 (2017)

R. Mrad, G. Chehab, Mechanical and microstructure properties of biochar-based mortar: an internal curing agent for PCC. Sustainability **11**(9) (2019). https://doi.org/10.3390/su11092491

Y. Nahata, K. Nirav, T.G. Tank, Effect of curing methods on efficiency of curing of cement mortar. APCBEE Proc. **9**, 222–229 (2014). https://doi.org/10.1016/j.apcbee.2014.01.040

D. Nduka, J. Ameh, J. Opeyemi, O. Rapheal, Awareness and benefits of self-curing concrete in construction projects: builders and civil engineers perceptions. Buildings **8**, 109 (2018). https://doi.org/10.3390/buildings8080109

A.M. Neville, *Properties of Concrete* (Pearson Education Limited, 2011)

S. Oh, C. Young, Superabsorbent polymers as internal curing agents in alkali activated slag mortars. Constr. Build. Mater. **159**, 1–8 (2018). https://doi.org/10.1016/j.conbuildmat.2017.10.121

L.J. Parrott, J.M. Illston, Load-Induced strains in hardened cement paste. J. Eng. Mech. Div. **101**, 13–24 (1975). https://doi.org/10.1061/jmcea3.0001986

R. Prommas, T. Rungsakthaweekul, Effect of microwave curing conditions on high strength concrete properties. Energy Proc. **56**, 26–34 (2014). https://doi.org/10.1016/j.egypro.2014.07.128

P. Rajappan, G.V.V.S.R. Kishore, C. Sundaramurthy, C.S. Pillai, A.K. Laharia, Effect of curing methods and environment on properties of concrete. Concr. Res. Lett. **5**, 786–811 (2014). https://doi.org/10.22214/ijraset.2018.5223

P. Rattanadecho, N. Makul, A. Pichaicherd, P. Chanamai, B. Rungroungdouyboon, A novel rapid microwave-thermal process for accelerated curing of concrete: prototype design, optimal process and experimental investigations. Constr. Build. Mater. **123**, 768–784 (2016). https://doi.org/10.1016/j.conbuildmat.2016.07.084

References

H.W. Reinhardt, S. Weber, Self-Cured high performance concrete introduction. J. Mater. Civ. Eng. **10**(November), 208–209 (1998). https://doi.org/10.1061/(asce)0899-1561(1997)9:1(1)

S. Riyazi, T.K. John, M. Matt, Super absorbent polymers (SAPs) as physical air entrainment in cement mortars. Constr. Build. Mater. **147**, 669–676 (2017). https://doi.org/10.1016/j.conbuildmat.2017.05.001

S.D. Rodríguez-Torres, N. Torres-Castellanos, Evaluation of Internal curing effects on concrete. Ingenieria e Investigacion (2019). https://doi.org/10.15446/ing.investig.v39n2.76505

D. Sarbapalli, Y. Dhabalia, K. Sarkar, B. Bhattacharjee, Application of SAP and PEG as curing agents for ordinary cement-based systems: impact on the early age properties of paste and mortar with water-to-cement ratio of 0.4 and above. Eur. J. Environ. Civ. Eng. **21**(10), 1237–1252 (2017). https://doi.org/10.1080/19648189.2016.1160843

G.K. Sastry, M.K. Putturu, Self-curing concrete with different self-curing agents. IOP Conf. Ser. Mater. Sci. Eng. **330**, 1–7 (2018). https://doi.org/10.1088/1757-899X/330/1/012120

T.R. Sathanandham, M. Gobinath, S. Naveenprabhu, K. Gnanasundar, S.G. Vajravel, R. Manoj, R. Jagathishprabu, Preliminary studies of self curing concrete with the addition of polyethylene glycol. Int. J. Eng. Res. Technol. **2**(11), 313–323 (2013)

E. Serri, S. Mohd Zailan, T. Roslan, R. Mahyuddin, Durability performance of oil palm shell lightweight concrete for insulation building material. Jurnal Teknologi **78**(5), 1–6 (2016). https://doi.org/10.11113/jt.v78.8228

S. Shahbazpanahi, M.K. Tajara, R.H. Faraj, A. Mosavi, Sustainable concrete containing recycled coarse aggregate. Crystal **11**(122), 15 (2021)

A. Shigeta, O. Yuko, K. Kenji, in *The 4th International Conference on Rehabilitation and Maintenance in Civil Engineering (ICRMCE 2018)*, vol 195. Microscopic Investigation on Concrete Cured Internally by Using Porous Ceramic Roof-Tile Waste Aggregate (2018), pp. 1–7

R.E. Spears, The 80 percent solution to inadequate curing problems. Concr. Int. **5**, 15–18 (1983)

Tang S., Y. Wang, Z. Geng, X. Xu, W.A. Yu, J. Chen, Structure, fractality, mechanics and durability of calcium silicate hydrates. Fract. Fract. **5**(2) (2021). https://doi.org/10.3390/fractalfract5020047

P.C. Taylor, Curing Concr. (2014). https://doi.org/10.1201/b15519

G. Thrinath, P.K. Sundara, Eco-friendly self-curing concrete incorporated with polyethylene glycol as self-curing agent. Int. J. Eng. Trans. A **30**(4), 473–478 (2017). https://doi.org/10.5829/idosi.ije.2017.30.04a.03

T. Uygunoğlu, I. Hocaoğlu, Effect of electrical curing application on setting time of concrete with different stress intensity. Constr. Build. Mater. **162**, 298–305 (2018). https://doi.org/10.1016/j.conbuildmat.2017.12.036

F.J. Vázquez-Rodríguez, N. Elizondo-Villareal, L. Hypatia Verástegui, A.T. Ana Maria, J. Fernando López-Perales, J. Eulalio Contreras de León, C. Gómez-Rodríguez, Effect of mineral aggregates and chemical admixtures as internal curing agents on the mechanical properties and durability of high-performance concrete. Materials **13**(9) (2020). https://doi.org/10.3390/ma13092090

R. Wasserman, A. Bentur, Efficiency of curing technologies: strength and durability. Mater. Struct. **46**, 1833–1842 (2013). https://doi.org/10.1617/s11527-013-0021-9

P.P. Woyciechowski, K. Maciej, The influence of dosing method and material characteristics of superabsorbent polymers (SAP) on the effectiveness of the concrete internal curing. Materials **11**, 1–21 (2018). https://doi.org/10.3390/ma11091600

L. Yang, X. Ma, J. Liu, X. Hu, Z. Wu, C. Shi, Improving the effectiveness of internal curing through engineering the pore structure of lightweight aggregates. Cem. Concr. Compos. **134**, 104773 (2022). https://doi.org/10.1016/j.cemconcomp.2022.104773

D. Ye, C.S. Shon, A.K. Mukhopadhyay, D.G. Zollinger, New performance-based approach to ensure quality curing during construction. J. Mater. Civil Eng. **22**, 687–695 (2010)

V.Z. Zadeh, P.B. Christopher, Nano-Mechanical properties of internally cured kenaf fiber reinforced concrete using nanoindentation. Cem. Concr. Compos. **52**, 9–17 (2014). https://doi.org/10.1016/j.cemconcomp.2014.04.002

B. Zhang, C.S. Poon, Internal curing effect of high volume furnace bottom ash (FBA) incorporation on lightweight aggregate concrete. J. Sustain. Cem. Based Mater. **6**(6), 366–383 (2017). https://doi.org/10.1080/21650373.2017.1299053

S. Zhutovsky, K. Konstantin, Influence of water to cement ratio on the efficiency of internal curing of high-performance concrete. Constr. Build. Mater. **144**, 311–316 (2017). https://doi.org/10.1016/j.conbuildmat.2017.03.203

S. Zhutovsky, K. Konstantin, B. Arnon, Revisiting the protected paste volume concept for internal curing of high-strength concretes. Cem. Concr. Res. **41**(9), 981–986 (2011). https://doi.org/10.1016/j.cemconres.2011.05.007

D. Zou, L. Kun, W. Li, H. Li, T. Cao, Effects of pore structure and water absorption on internal uring efficiency of porous aggregates. Constr. Build. Mater. **163**, 949–959 (2018). https://doi.org/10.1016/j.conbuildmat.2017.12.170

Chapter 2
Mechanisms of Conventional and Self Cured Cement Hydration

2.1 Introduction

This chapter provides an exploration of the mechanisms behind both conventional and self cured cement hydration. It delves into the intricate series of chemical reactions that occur during the transformation of dry, powdery cement into the solid and robust material of concrete. Initially, the chapter lays out the fundamental principles of the cement hydration process, offering an explanation of the mechanisms at play in the hydration of conventionally cured concrete. Following this, it examines the various factors that influence the curing process, highlighting the differences and similarities between conventional curing and self curing methods in cement-based systems. Consequently, the significance of self curing concrete including an analysis of the environmental conditions and material properties that impact the efficiency and effectiveness of self curing processes was discussed.

2.2 Cement Hydration of Conventional Concrete

Cement hydration is the chemical reaction between cement particles and water that leads to the hardening of concrete. This process is a combination of dissolution, chemical reactions and the formation of crystalline structures, which together develop the mechanical strength and durability of concrete. When water is added to cement, it initiates a series of hydration reactions that gradually convert the mixture into a hardened material. This transformation is not merely a drying process but involves complex chemical changes that create a microstructure capable of bearing loads and resisting environmental factors.

Portland cement, the most widely used type of cement, is composed of several mineral compounds, each playing a crucial role in the hydration process. The primary components are:

- **Tricalcium Silicate (C_3S)**: This compound hydrates and hardens rapidly, contributing significantly to the early strength of concrete.
- **Dicalcium Silicate (C_2S)**: It hydrates more slowly, contributing to the long-term strength of concrete.
- **Tricalcium Aluminate (C_3A)**: This compound reacts quickly with water and significantly influences the setting time and early strength of the concrete.
- **Tetracalcium Aluminoferrite (C_4AF)**: It reacts moderately quickly and contributes to the strength and color of the cement.
- **Gypsum ($CaSO_4 \cdot 2H_2O$)**: It is added to control the setting time of cement.

These compounds, in combination with water, undergo a series of chemical reactions that result in the formation of new products and the hardening of the mixture which will be explained in Sect. 2.3.

The addition of water to OPC powder initiates the cement hydration reactions instantly. This series of chemical reactions leads in the cement paste setting and hardening. Within a few minutes, needle-like crystals of calcium sulfoaluminate hydrate, notably ettringite, develop. After a period of time, ettringite converts into monosulfate hydrate. Two (2) hours after the cementation process begins, large prismatic crystals of calcium hydroxide (C–H) and tiny calcium silicate hydrates (C–S–H) fill the voids formerly filled by water and hydrated cement particles. Therefore, calcium silicate hydrate, calcium hydroxide and calcium sulfoaluminate are the three primary components of hydrated cement paste (Saleh and Eskander 2020). C–S–H is the primary hydration product, contributing to almost 60% of the volume of solids. It is composed of a layer of sponge-like structures with a huge surface area (500 m^2/g).

The ultimate strength of the product is largely owing to the development of C–S–H and is principally due to van der Waals physical adhesion forces. C–H is the second most prevalent component, contributing to approximately 25% of the total. Compared to C–S–H, it is composed of massive plate-like crystals with a lower surface area. It contributes to the reduction of van der Waal forces and is relatively soluble in comparison to C–S–H, making the concrete reactive to acidic solutions. Calcium sulfoaluminate plays a small part in the cementitious structure properties by almost 15% solid volume. Chemical resistance of the cementitious final product to sulfate attack is an issue, owing to the existence of the monosulfate hydrate (Saleh and Eskander 2020).

Figure 2.1 shows electron microscopic images of hardened cement paste after hydration. The main reactions that occur when mixing water and gypsum with Portland cement powder are given in the following equations:

$$2\underset{\text{Alite}}{Ca_3SiO_5} + 7H_2O \leftrightarrow \underset{\text{C–S–H}}{Ca_3Si_2O_3(OH)_8} + 3\underset{\text{portlandite}}{Ca(OH)_2} \quad (2.1)$$

$$2\underset{\text{Belite}}{Ca_2SiO_2} + 5H_2O \leftrightarrow \underset{\text{C–S–H}}{Ca_3Si_2O_3(OH)_8} + \underset{\text{portlandite}}{Ca(OH)_2} \quad (2.2)$$

2.2 Cement Hydration of Conventional Concrete

$$Ca_4Al_2Fe_2O_{10} + 7H_2O \leftrightarrow \underset{\text{Hydrogarnet}}{Fe_2O_3 + Ca_3Al_2(OH)_{12}} + \underset{\text{portlandite}}{Ca(OH)_2} \quad (2.3)$$

$$\underset{\text{Portlandite}}{CaO_4 + H_2O} \leftrightarrow Ca(OH)_2 \quad (2.4)$$

$$\underset{\text{Aluminate}}{Ca_3Al_2O_6} + \underset{\text{Gypsum}}{3\ CaSO_4 2H_2O} + 26\ H_2O$$

$$\leftrightarrow \left[\underset{\text{Ettringite}}{Ca_3Al(OH)_6 \cdot 12H_2O}\right]2(SO_4) \cdot 32H_2O \quad (2.5)$$

$$\left[\underset{\text{Ettringite}}{Ca_3Al(OH)_6 \cdot 12H_2O}\right]2(SO4)_3 \cdot 2H_2O + \underset{\text{Aluminate}}{Ca_3Al_2O_6} + 4\ H_2O$$

$$\leftrightarrow 3\underset{\text{Monosulfate}}{\left[Ca_2Al(OH)_6 \cdot 2H_2O\right]_2 SO_4 \cdot 2H_2O} \quad (2.6)$$

Fig. 2.1 Electron microscopic images of hardened cement paste after hydration (Stutzman 2001; used with permission of John Wiley and Sons, from Materials Science of Concrete, Special Volume: Calcium Hydroxide in Concrete, Copyright 2001; permission conveyed through Copyright Clearance Center, Inc.)

2.3 Stages of Hydration Process

The hydration of cement can be divided into several stages, each is characterized by distinct chemical reactions and changes in the material's properties. These stages include:

2.3.1 Initial Mixing and Pre-induction Period

When water is first added to cement, the surface of the cement particles becomes wet and some dissolution of the components occurs. This initial phase, known as the pre-induction period, lasts only a few minutes. During this time, a thin layer of hydration products begins to form on the surface of the cement particles.

2.3.2 Induction Period

Following the initial wetting, the hydration process enters a relatively dormant phase called the induction period. This stage can last from a few hours to several hours, depending on various factors like temperature and the presence of admixtures. During the induction period, the hydration products formed during the pre-induction period act as a barrier, slowing down further reaction.

2.3.3 Acceleration Period

The induction period is followed by the acceleration period, during which the rate of hydration increases significantly. This stage is characterized by the rapid formation of calcium silicate hydrate (C–S–H) and calcium hydroxide (C–H), which are crucial for the development of strength in the cement paste.

During the acceleration period, tricalcium silicate (C_3S) and dicalcium silicate (C_2S) react with water to form C–S–H and C–H and the reactions are as follows:

- **Tricalcium Silicate (C_3S)**

  The chemical formula for tricalcium silicate (C_3S) is Ca_3SiO_5. When tricalcium silicate (C_3S) hydrates, it reacts with water (H_2O) to form calcium silicate hydrate (C–S–H) and calcium hydroxide ($Ca(OH)_2$). The equation for this reaction can be written as:

  $$2Ca_3SiO_5 + 7H_2O \rightarrow 3CaO \cdot 2SiO_2 \cdot 4H_2O + 3Ca(OH)_2$$

2.4 Microstructural Development of Hydration Products

The products from this reaction are calcium silicate hydrate (C–S–H) and calcium hydroxide (Ca(OH)$_2$). This reaction is exothermic and releases heat and contributes to the early strength development of concrete.

- **Dicalcium Silicate (C$_2$S)**

The chemical formula for dicalcium silicate (C$_2$S) is Ca$_2$SiO$_4$. Similar to tricalcium silicate, dicalcium silicate also forms calcium silicate hydrate (C–S–H) and calcium hydroxide (C–H) during hydration. When dicalcium silicate (C$_2$S) hydrates, it reacts with water (H$_2$O) to form calcium silicate hydrate (C–S–H) and calcium hydroxide (Ca(OH)$_2$), similar to tricalcium silicate but at a slower rate and with less heat generation. The equation for this reaction can be written as follows:

$$2Ca_2SiO_4 + 5H_2O \rightarrow 3CaO \cdot 2SiO \cdot 4H_2O + Ca(OH)_2$$

Calcium silicate hydrate (C–S–H) is the primary hydrate phase responsible for binding cement particles together and providing strength. Calcium hydroxide (Ca(OH)$_2$) is a by-product of hydration, which contributes to the alkali content in concrete and can affect its durability over time.

The hydration reactions proceed rapidly during the initial stages after mixing cement with water, leading to the early setting and hardening of concrete.

The C–S–H gel, a complex amorphous structure, is the primary binding phase in concrete, contributing significantly to its strength and durability. The calcium hydroxide (C–H), on the other hand, is a crystalline phase that fills the pores and contributes to the material's pH stability.

2.3.4 Deceleration and Steady-State Periods

As the hydration process continues, the rate of reaction gradually decreases, entering the deceleration period. The formation of hydration products slows down as the available water and unreacted cement diminish. Eventually, the process reaches a steady-state period, where hydration continues at a much slower rate, leading to the long-term development of concrete strength.

2.4 Microstructural Development of Hydration Products

The microstructure of hydrated cement paste plays a critical role in determining the properties of concrete. The development of this microstructure is a complex process influenced by the hydration reactions and the spatial arrangement of the hydration products which will be elaborated in the following sub-sections.

2.4.1 Formation of C–S–H Gel

The C–S–H gel, formed primarily from the hydration of C_3S and C_2S, is the most significant phase in the hardened cement paste. It is a nanostructured material with a highly complex and variable composition. The C–S–H gel binds the aggregate particles together, providing the bulk of the mechanical strength and contributing to the material's durability.

2.4.2 Growth of C–H Crystals

C–H also known as portlandite, crystallizes in the pores of the cement paste. These crystals are relatively large and well-defined, providing additional strength and helping to maintain the high pH necessary for protecting embedded steel reinforcement from corrosion.

2.4.3 Formation of Ettringite and Monosulfoaluminate

Tricalcium aluminate (C_3A) reacts with gypsum and water to form ettringite during the early stages of hydration and the reaction is as follows:

$$C_3A + 3CaSO_4 \cdot 2H_2O + 26H_2O \rightarrow C_6AS_3 \cdot 32H_2O$$

where $C_6AS_3 \cdot 32H_2O$ or $Ca_6 Al_2 (SO4)_3 (OH)_{12} \cdot 26H_2O$ is ettringite.

As hydration progresses, ettringite may react further with additional C_3A to form monosulfoaluminate. Ettringite ($Ca_6 Al_2 (SO4)_3 (OH)_{12} \cdot 26H_2O$) can react with additional tricalcium aluminate (C_3A) to form monosulfoaluminate (also known as calcium aluminate monosulfate or AFm phase). This secondary reaction typically occurs after the initial formation of ettringite when the supply of gypsum is depleted.

$$C_6AS_3 \cdot 32H_2O + 2C_3A + 4H_2O \rightarrow 3C_4AS \cdot 18H_2O$$

Monosulfoaluminate (or $Ca_4Al_2(SO_4)(OH)_{12} \cdot 6H_2O$ is a less expansive and stable phase compared to ettringite, contributing to the final hardened structure of cement. These phases contribute to the setting and early strength of the cement paste, as well as influencing its dimensional stability and resistance to sulfate attack.

2.5 Factors Influencing Hydration

2.4.4 Pore Structure and Capillary Porosity

The hydration process significantly alters the pore structure of the cement paste. Initially, the paste is a highly porous system with capillary pores filled with water. As hydration progresses, these pores become filled with hydration products, reducing the overall porosity and increasing the material's density.

The size and distribution of the remaining capillary pores are crucial factors influencing the permeability, durability and strength of the concrete. A well-hydrated paste with low capillary porosity will exhibit high strength and durability, while a paste with significant capillary porosity may be more susceptible to cracking and degradation.

2.5 Factors Influencing Hydration

Several factors can influence the rate and extent of cement hydration, affecting the properties of the resulting concrete. These factors include:

2.5.1 Water-Cement Ratio (W/C)

The water-cement ratio is a critical parameter in concrete mix design. Cement hydration involves a series of chemical reactions between the cement particles and water as explained in earlier section. These reactions form various hydration products, including calcium silicate hydrate (C–S–H) and calcium hydroxide (C–H), which are responsible for the strength and durability of the concrete. The W/C ratio determines the availability of water for these reactions. Adequate water is needed to fully hydrate the cement particles, leading to the formation of a dense and strong matrix.

W/C also affects the workability of the fresh concrete and the porosity of the hardened paste. A lower water-cement ratio generally leads to higher strength and durability, as it reduces the capillary porosity and promotes the formation of a dense microstructure.

2.5.2 Temperature

Temperature plays a significant role in the rate of cement hydration. Higher temperatures accelerate the hydration reactions, leading to faster strength development but potentially causing thermal stresses and cracking. Conversely, lower temperatures slow down the hydration process, which can be beneficial in controlling heat generation and reducing the risk of early-age cracking.

2.5.3 Cement Composition

The chemical composition of cement influences its hydration behavior and the properties of the resulting concrete. For example, high C_3S content promotes rapid early strength development, while higher C_2S content contributes to long-term strength gain. The presence of other components, such as gypsum and various admixtures, can also modify the hydration process and its outcomes.

2.5.4 Admixtures

Admixtures are added to the concrete mix to enhance specific properties or control certain aspects of the hydration process. For example, superplasticisers improve workability without increasing the water-cement ratio, while retarders delay the setting time to allow for longer working periods. These additives can significantly influence the rate and extent of hydration, affecting the final properties of the concrete.

2.6 Mechanisms of Self Curing Concrete

Self curing technology or internal curing has gained popularity within concrete research community. The concept of self curing concrete is to mitigate water evaporation in concrete and hence increase the capacity of water retention in concrete (Wang et al. 1994; Bentz et al. 2005; Dhir et al. 1998; El-Dieb 2007; Bora et al. 2017; Akeed et al. 2022). Therefore, the technique has been introduced whereby water-filled internal curing agents react as reservoirs are added to concrete mixture which can gradually release its water during hydration and evaporation process (Bentz and Weiss 2011; Liu et al. 2017; Jongvisuttisun et al. 2018; Sastry and Kumar 2018; Yang et al. 2022) as illustrated in Fig. 2.2. High-performance concrete (HPC) mixtures were developed because of the increasing issues on concrete durability (Hoff 2002; Babcock and Taylor 2015; Akeed et al. 2022) and due to the use of lower water-cementitious (w/c) material ratios, as well as chemical admixtures and supplementary cementitious material (SCMs). However, well-hydrated in the cementitious system is an indicator of achieving such performance by properly cured (Meeks and Carino 1999). Low W/C ratio concrete mixtures, which is less than 0.42, are unable to fully hydrate the cement in the mixture due to insufficient water (Neville 2011). Therefore, the benefit of self curing concrete from absorbed moisture in porous aggregate was identified to solve the problem of insufficient water in HPC mixtures by providing extra water to replace that depleted during the process of cement hydration.

Porous aggregate is frequently associated with poor concrete quality. However, when utilized in wet conditions, the aggregate might benefit the concrete since the water absorbed by the aggregate is slowly released into the already-hardened

2.6 Mechanisms of Self Curing Concrete

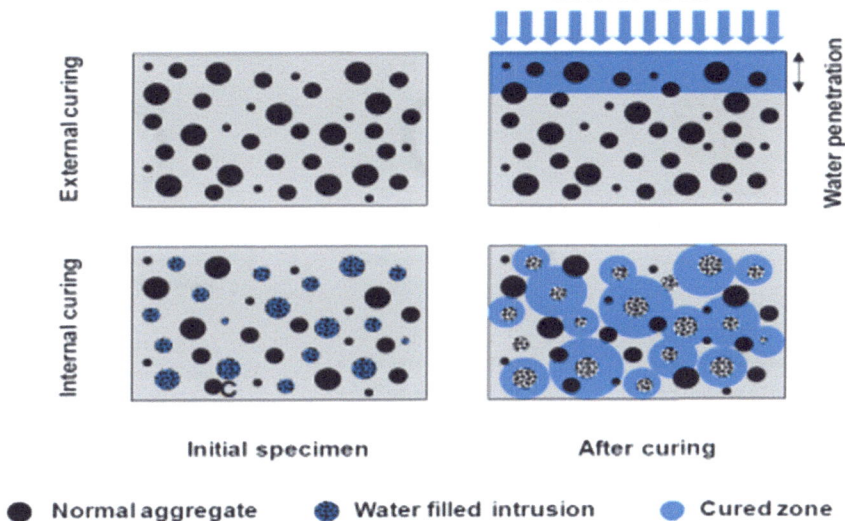

Fig. 2.2 Illustration of the differences between self curing and external curing (Bentz and Weiss 2011)

cement paste, continuing the hydration process (Sampebulu' 2012). As a result, concrete properties such as increased strength and decreased drying shrinkage will be improved. The water movement is caused by the humidity gradient between the aggregate that is high and the cement paste that is low. Figure 2.3 shows the schematic diagram of the mechanism of self curing concrete.

Self curing also known as autogenous curing or internal curing, enables curing "from the inside out," which is achieved by introducing a pre-saturated component as an internal curing agent (Han et al. 2017). The curing agent is spread uniformly throughout the matrix and acts as a reservoir for internal water. The water within the curing agent has not involved in the chemical reaction until a humidity gradient forms during an initial hydration phase. Han et al. (2017) illustrate the self curing process occurs as shown in Fig. 2.4. Water is transported from the curing agent to unhydrated cement by the driving forces of capillary suction, vapor diffusion and capillary condensation for supporting continuous hydration. As a result, chemical shrinkage and self-desiccation due to low water binder ratio (w/b) can be significantly reduced.

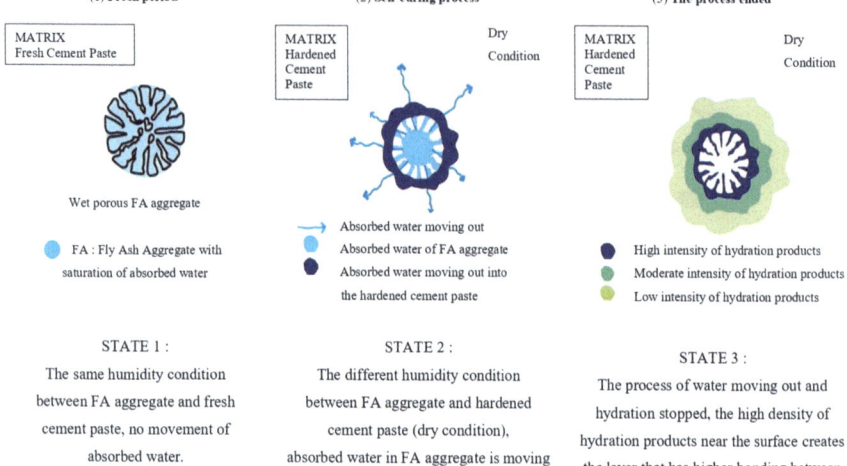

Fig. 2.3 Mechanism of self curing concrete (Sampebulu' 2012; licensed under a Creative Commons Attribution (CC BY))

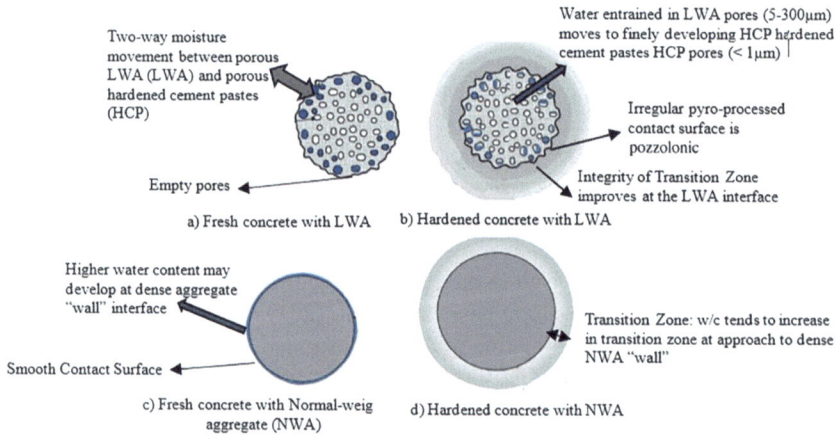

Fig. 2.4 Contact zone under internal curing and normal curing (Han et al. 2017; Reproduced with permission from Springer Nature)

2.7 Significance of Self Curing Concrete

The scope of self curing concrete extends beyond mere technological innovation; it embodies a paradigm shift toward sustainable construction practices. By reducing reliance on water-intensive curing methods and enhancing concrete performance through internal moisture management, self curing agents contribute to resource conservation, energy efficiency and environmental stewardship. The significance of self curing concrete lies not only in its potential to improve construction efficiency and durability but also in its alignment with global sustainability goals and resilience in the face of changing environmental conditions.

In addition, the utilization of natural aggregates has become a growing issue, given the excessive use for construction purposes. Recycling and using by-products have attracted growing interest and attention from researchers to reduce wastage. As such, the development of new methods in managing waste has swiftly become one of the most important research topics. The increasing need to reuse materials, given diminishing natural resources, has become a recognized and volatile issue healthy debated by scholars and researchers. Accordingly, research studies have increasingly investigated artificial aggregates for infrastructure construction as a replacement for natural aggregates. Most artificial aggregates produced from waste or by-products (Al Saffar et al. 2019; Safiuddin et al. 2013; Xiao et al. 2012; Tabsh and Abdelfatah 2009) and their unique properties, such as porosity and absorption capacity, have the added benefits of supplying water internally for self curing concrete (Shigeta et al. 2018; Wyrzykowski et al. 2016; Mousa et al. 2015). In addition, using large quantities of industrial and agricultural waste can help to reduce suspension in natural aggregates (He et al. 2020; Shafigh et al. 2014). Moreover, it is seen as an effective process to convert waste or by-products into valuable building materials. Thus, this leads to beneficial performance for the economy whereby it produces cheaper mortar and concrete materials for low-cost construction (Al Saffar et al. 2019). The utilization of artificial aggregates produced from waste materials will also aid in the reduction and usage of materials made from chemical resources as curing agents in concrete. Furthermore, it positively contributes to safeguarding and protecting the environment leading to sustainable development and reduces carbon emission (He et al. 2020). Besides, waste material as a curing agent replacing natural aggregates in concrete will lead to the reduction in landfills and prevent natural resources such as flora and fauna from being destroyed (Tabsh and Abdelfatah 2009).

The review of research on self curing concrete revealed it to be both dense and durable compared to conventional non-cured concrete (Chand et al. 2018; Ma et al. 2019; El-Dieb et al. 2012), thus increasing the building's lifespan (i.e., service life). The growing scarcity of water resources in hot climates, such as in Afro-Asian regions, has also required regular checking on the use of freshwater for concrete production, given the approximate rate of $3 m^3$ for every $1 m^3$ of concrete produced for curing purposes (Sampebulu'2012; Sarbapalli et al. 2016). Therefore, the techniques for self curing concrete need to be monitored and advanced, where possible, by conserving the use of freshwater. For example, unneeded water spraying or water sprinkling techniques are needless and should be avoided. Therefore, the self curing curing agent provides an avenue for further cement hydration via the absorption of water before or during the mixing of concrete as the absorbed water can be slowly released during the process of hardened concrete (Liu et al. 2017; Ma et al. 2019).

2.8 Future Studies in Self Curing Concrete

Understanding the mechanisms of cement hydration is fundamental to controlling and optimizing the properties of concrete. The complex interplay of chemical reactions, microstructural development and external factors determines the performance of the hardened material. In the context of self curing concrete, insights into conventional hydration processes are essential for appreciating how innovative technologies can enhance and sustain hydration, leading to improved durability and performance.

By delving into the intricacies of cement hydration, the groundwork is laid for exploring advanced concrete technologies, such as self curing systems, which promise to revolutionize the field of construction.

To fully harness the potential of self cured concrete, these future research directions include technological advancements, interdisciplinary approaches and policy support to facilitate the adoption of these sustainable practices which are listed below:

1. **Developing Advanced Self Curing Agents:** Research into new or enhanced materials that can serve as internal curing agents, particularly those derived from sustainable sources, is crucial. This could involve exploring nanotechnology, bio-based materials, or advanced polymers that provide superior water retention and release properties.
2. **Optimizing Green Aggregate Properties:** Investigating ways to improve the performance of GAA, such as enhancing their strength, durability or compatibility with self curing agents, will be essential. This includes studying the impact of aggregate properties on the overall concrete mix and developing methods to standardize their use.
3. **Performance Evaluation and Long-Term Studies:**
    i. **Comprehensive Performance Assessment** Conducting extensive studies to evaluate the long-term performance of self cured concrete with green aggregates under various environmental and loading conditions is necessary. This includes understanding how these materials behave over time and ensuring they meet or exceed the performance of traditional concrete.
    ii. **Lifecycle Analysis and Sustainability Metrics:** Implementing rigorous lifecycle analysis to quantify the environmental benefits of using self cured concrete with green aggregates is vital. This includes assessing the carbon footprint, resource consumption and waste reduction associated with these materials throughout their lifecycle.

References

M.H. Akeed, S. Qaidi, R.H. Faraj, A.S. Mohammed, W. Emad, B.A. Tayeh, A.R.G. Azevedo, Ultra-high-performance fiber-reinforced concrete. Part II: hydration and microstructure. Case Stud. Constr. Mater. **17**(June), e01289 (2022). https://doi.org/10.1016/j.cscm.2022.e01289

D.M. Al Saffar, A.J.K. Al Saad, B.A. Tayeh,Effect of internal curing on behavior of high performance concrete: an overview. Case Stud. Constr. Mater. **10**, e00229 (2019). https://doi.org/10.1016/j.cscm.2019.e00229.

A.E. Babcock, P. Taylor,Impacts of internal curing on concrete properties. Eng. Mater. Sci. (2015). https://api.semanticscholar.org/CorpusID136853760

D.P. Bentz, P. Lura, J.W. Roberts, Mixture proportioning for internal curing. Concr. Int. **27**(2), 35–40 (2005)

D.P. Bentz, W.J. Weiss, *Internal curing : A 2010 State-of-the-Art Review. U.S. Department of Commerce, National Institute of Standards and Technology* (2011)

M. Bora, M. Vohra, P.M. Sakil, D. Vyas,Self-curing concrete-literature review. Int. J. Eng. Dev. Res. **5**. www.ijedr.org

M.S.R. Chand, P.R. Kumar, P.S.N.R. Giri, G.R. Kumar, Performance and microstructure characteristics of self-curing self-compacting concrete. Adv. Cem. Res. **30**, 451–468 (2018). https://doi.org/10.1680/jadcr.17.00154

R.K. Dhir, P.C. Hewlett, T.D. Dyer, Mechanisms of water retention in cement pastes containing a self-curing agent. Mag. Concr. Res. **50**(1), 85–90 (1998)

A.S. El-Dieb, Self-curing concrete : Water retention, hydration and moisture transport. Constr. Build. Mater. **21**, 1282–1287 (2007). https://doi.org/10.1016/j.conbuildmat.2006.02.007

A.S. El-Dieb, T.A. El-Maaddawy, A.A.M. Mahmoud, Water-soluble polymers as selfcuring agents in cement mixes. Adv. Cem. Res. **24**(5), 291–299 (2012). https://doi.org/10.1680/adcr.11.00030

B. Han, L. Zhang, J. Ou, in *Smart and Multifunctional Concrete Toward Sustainable Infrastructures* (2017), pp. 1–400. https://doi.org/10.1007/978-981-10-4349-9

J. He, S. Kawasaki, V. Achal, The utilization of agricultural waste as agro-cement in concrete: a review. Sustainability **12**(17), 16 (2020). https://doi.org/10.3390/su12176971

G.C. Hoff,The use of lightweight fines for the internal curing of concrete. Rep. Northeast Solite Corp. 1–44 (2002)

P. Jongvisuttisun, J. Leisen, K.E. Kurtis, Key mechanisms controlling internal curing performance of natural fibers. Cem. Concr. Res. **107**, 206–220 (2018). https://doi.org/10.1016/j.cemconres.2018.02.007

F. Liu, J. Wang, X. Qian, J. Hollingsworth, Internal curing of high performance concrete using cenospheres. Cem. Concr. Res. **95**, 39–46 (2017). https://doi.org/10.1016/j.cemconres.2017.02.023

X. Ma, J. Liu, C. Shi, A review on the use of LWA as an internal curing agent of high performance cement-based materials. Constr. Build. Mater. **218**, 385–393 (2019). https://doi.org/10.1016/j.conbuildmat.2019.05.126

K.W. Meeks, N.J. Carino, *Curing of High-Performance Concrete: Report of the State-of-the-Art*. National Institute of Standards and Technology (1999)

M.I. Mousa, M.G. Mahdy, A.H. Abdel-reheem, A.Z. Yehia, Self-curing concrete types: Water retention and durability. Alex. Eng. J. **54**(3), 565–575 (2015). https://doi.org/10.1016/j.aej.2015.03.027

A.M. Neville, *Properties of Concrete*. Pearson Education Limited (2011)

M. Safiuddin, U.J. Alengaram, M.M. Rahman, M.A. Salam, M.Z. Jumaat, Use of recycled concrete aggregate in concrete: a review. J. Civ. Eng. Manag. **19**(6), 796–810 (2013). https://doi.org/10.3846/13923730.2013.799093

H.M. Saleh, S.B. Eskander, in *New Materials in Civil Engineering*. Innovative Cement-Based Materials for Environmental Protection and Restoration (INC, 2020), pp. 613–641. https://doi.org/10.1016/b978-0-12-818961-0.00018-1

V. Sampebulu', Increase on strengths of hot weather concrete by self-curing of wet porous aggregate. Civil Eng. Dimens. **14**(2) (2012).

D. Sarbapalli, Y. Dhabalia, K. Sarkar, B. Bhattacharjee, Application of SAP and PEG as curing agents for ordinary cement- based systems: impact on the early age properties of paste and mortar with water-to-cement ratio of 0.4 and above. Eur. J. Environ. Civil Eng. **21**, 1237–1252 (2016). https://doi.org/10.1080/19648189.2016.1160843

G.K. Sastry, P.M. Kumar,Self-curing concrete with different self-curing agents. IOP Conf. Ser. Mater. Sci. Eng. **330**, 1–7 (2018). https://doi.org/10.1088/1757-899X/330/1/012120

P. Shafigh, H. Mahmud, M.Z. Jumaat, M. Zargar, Agricultural wastes as aggregate in concrete mixtures—a review. Constr. Build. Mater. **53**, 110–117 (2014). https://doi.org/10.1016/j.conbuildmat.2013.11.074

A. Shigeta, Y. Ogawa, K. Kawai, in *The 4th International Conference on Rehabilitation and Maintenance in Civil Engineering (ICRMCE 2018)*, vol. 195. Microscopic Investigation on Concrete Cured Internally by Using Porous Ceramic Roof-Tile Waste Aggregate (2018), pp. 1–7

P.E. Stutzman, in *Calcium Hydroxide in Concrete Proceedings*, ed. by J. Skalny, J. Gebauer, I. Odler. Scanning Electron Microscopy in Concrete Petrograph (2001), pp. 59–72

S.W. Tabsh, A.S. Abdelfatah, Influence of recycled concrete aggregates on strength properties of concrete. Constr. Build. Mater. **23**(2), 1163–1167 (2009). https://doi.org/10.1016/j.conbuildmat.2008.06.007

J. Wang, R.K. Dhir, M. Levitt, Membrane curing of concrete: moisture loss. Cem. Concr. Res. **24**(8), 1463–1474 (1994). https://doi.org/10.1016/0008-8846(94)90160-0

M. Wyrzykowski, S. Ghourchian, S. Sinthupinyo, N. Chitvoranund, T. Chintana, P. Lura, Internal curing of high performance mortars with bottom ash. Cem. Concr. Compos. **21**, 1–9 (2016). https://doi.org/10.1016/j.cemconcomp.2016.04.009

J. Xiao, W. Li, Y. Fan, X. Huang, An overview of study on recycled aggregate concrete in China (1996–2011). Constr. Build. Mater. **31**, 364–383 (2012). https://doi.org/10.1016/j.conbuildmat.2011.12.074

L. Yang, X. Ma, J. Liu, X. Hu, Z. Wu, C. Shi,Improving the effectiveness of internal curing through engineering the pore structure of lightweight aggregates. Cem. Concr. Compos. 104773 (2022). https://doi.org/10.1016/j.cemconcomp.2022.104773

Chapter 3
Types of Self Curing Agents and Properties of Self Cured Concrete/Mortar

3.1 Introduction

This chapter delves into the innovative realm of self curing agents, which are revolutionizing the field of concrete and mortar technology. Self curing, or internal curing, represents a significant advancement, allowing concrete and mortar to retain moisture over extended periods, thereby enhancing their hydration and overall performance. This chapter begins by exploring the various types of self curing agents, including artificial lightweight aggregates, normal weight aggregates, porous superfine powders, hydrophilic polymers chemical admixtures and natural fibers, each with unique mechanisms for facilitating internal curing. Following this, the effects of these agents impart on the physical, mechanical and durability properties namely workability, density, compressive, tensile and flexural strength of concrete and mortar are examined. This chapter also provides an understanding of how integrating self curing agents can optimize concrete and mortar's performance.

3.2 Types of Self Curing Agents and Mechanism in Cementitious Materials

There are a lot of techniques that have been studied in self curing concrete by using different materials of curing agent. The techniques are namely lightweight aggregate (Akhnoukh 2017; Zhutovsky and Kovler 2017; Kevern and Nowasell 2018; Kim and Lee 2018; Zou et al. 2018), porous superfine powders (Ye et al. 2013; Van et al. 2014; Liu et al. 2017; Al Saffar et al. 2019), superabsorbent polymers (Riyazi et al. 2017; Almeida and Klemm 2018; Kang et al. 2018; Oh and Cheol 2018; Woyciechowski and Kalinowski 2018), polyethylene glycol (Bashandy 2016; Madduru et al. 2016, 2018; Sarbapalli et al. 2017), natural fibers (Kawashima and Shah 2011; Zadeh and Bobko 2014; Jongvisuttisun et al. 2018) and normal weight aggregate (Abate et al.

2018; Kon et al. 2018; Shigeta et al. 2018) have also been looked into as curing agents. The description for each curing agent and their mechanisms lie behind in self curing concrete is reviewed in the following sub-sections.

3.2.1 Artificial Lightweight Aggregate

The most widely utilized self curing agent in self curing concrete is lightweight aggregate (LWA). Artificial LWA is manufactured from waste materials like as fly ash, expanded clay, slate, shale, bottom ash, etc. The problem of disposing industrial by-products, such as fly ash, silica fume and bottom ash, grows daily. As an environmentally friendly material for the construction industry, the ideal lightweight aggregate should have a sintered core with a nearly spherical shape (diameter in the range of 4 to14 mm) and a rough surface that is impervious to water, as well as strong features such as low porosity and water absorption (Wang et al. 2020). Typically, the density of concrete ranges from 2200 to 2600 kg/m^3. Regardless of those, LWA can provide as self curing agent in the concrete as reported by Balapour et al. 2020; Pradeep and Beena 2020; Ma et al. 2019.

Pre-wetted lightweight aggregates immersed in water at least for 1 h have often been used as internal reservoirs which a system of capillary pores in cement paste is formed during hydration. As soon as the relative humidity (RH) decreases (due to hydration and drying), a humidity gradient develops (Hoff 2002; Grasley et al. 2006; Zhang et al. 2012; Yadav et al. 2017). The migration of water in concrete based on the law of fluid flow and the system's law of capillary attraction is illustrated in Fig. 3.1.

As observed in the figure, the radius of pores in cement paste (r(t)) is smaller than the pores in LWA (Ra). The pores of the cement paste by capillary suction absorbs the

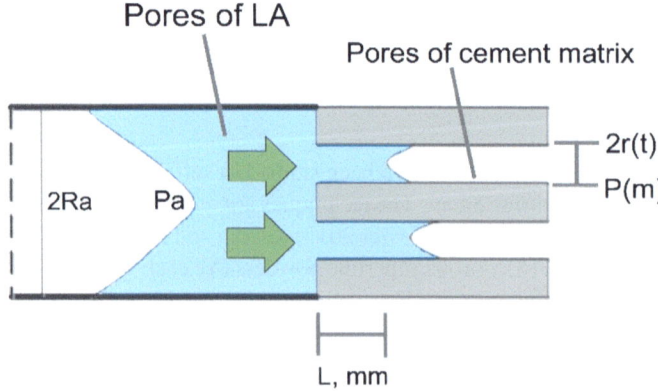

Fig. 3.1 Model moving of water in concrete incorporate self curing with $r(t) <$ Ra (Nguyen and Le 2018; licensed under a Creative Commons Attribution (CC BY))

water from the LWA due to difference in vapor pressure and transports the water to the drier cement paste, where a reaction with the un-hydrated cement occurs (Weber and Reinhardt 1997; Hoff 2002; Yadav et al. 2017; Nguyen and Le 2018). The un-hydrated cement particles, hydrated to form hydration products, reduce the size of the pores, enabling the pores to continue absorbing the water from the LWA. This process continues until all the water from LWA has been transported to the cement paste creating self curing mechanism.

Among the various features found in LWA, it has been shown to have high porosity and absorption capacity with the added benefit of supplying curing water internally (Paul and Lopez 2011; de Sensale and Goncalves 2014; Akhnoukh 2017; Bari et al. 2021). Moreover, due to contain of porosity, the density of LWA is less than normal-weight aggregates. The saturated-surface dried density of LWA is reported to be below than 2000 kg/m^3 (Zhutovsky et al. 2002a, b; Akcay and Tasdemir 2010; Paul and Lopez 2011; Kim et al. 2016; Zhang and Poon 2017; Zou et al. 2018). However, water absorption of porous LWA depend on the pore structures. The minimum water absorption above 5% by mass of LWA is required as recommended by ASTM C1761 /C1761M (Castro et al. 2011; Lura et al. 2014; ASTM C1761 2017). LWA can be in the form of coarse and fine aggregate. The most common ones used as curing agent in self curing concrete from natural-based are expanded shale (Zou and Weiss 2014; Yang and Wang 2017; Yang et al. 2022), expanded clay (LECA) (Paul and Lopez 2011; Ghourchian et al. 2013; Mousa et al. 2015a, b; Akhnoukh 2017; Zhang and Poon 2017; Kamal et al. 2018; Al Saffar et al. 2019; Agrawal et al. 2021) and pumice (Semion et al. 2002; Mustafa et al. 2009; Akcay and Tasdemir 2010; Abadel 2023). Meanwhile, curing agent from by-product materials which often studied in concrete by several researchers are bottom ash (Kim et al. 2014, 2016; Wyrzykowski et al. 2016; Mohamad et al. 2017; Zhang and Poon 2017; Kim and Lee 2018; Nguyen et al. 2019), lightweight concrete waste (Suwan and Wattanachai 2017) and brick chips (Iffat et al. 2017; Manzur et al. 2019; Bandara et al. 2019).

3.2.2 Porous Superfine Powders

Porous superfine powders are small particle size materials with a large specific surface area and a mesoporous structure. It can absorb the aqueous phase, enabling water supply for the hydration process in the cementitious material (Rahmasari et al. 2019; Liu et al. 2017). Porous superfine powders possess nanometer-size of pores, for instance, cenosphere (Wang and Liu 2016; Liu et al. 2017, 2019; Chen et al. 2019), rice husk ash (Tuan et al. 2011; Ye et al. 2013; Van et al. 2014; Amin et al. 2019) and biochar (Gupta and Kua 2018; Mrad and Chehab 2019). Generally, the particle size of porous superfine powders ranges between 5 and 10 μm, which is much smaller compared to SAP particles and LWA, where the pore size ranges between 4 and 10 nm as illustrated in Fig. 3.2. Only if the pore sizes of cement paste smaller than superfine powders' pore size will the water be released into cement paste, in which the hydration process occurs.

Fig. 3.2 Microstructure of cenopheres under SEM observation (Liu et al. 2019; licensed under a Creative Commons Attribution (CC BY))

In addition, the application of porous superfine powders capable of slowing down the internal RH (self-desiccation) in Ultra High Performance Concrete (UPHC) significantly reduces its autogenous shrinkage (de Sensale et al. 2008; Van et al. 2014; Liu et al. 2017). According to Kelvin equation (Lura et al. 2003), the capacity of water saturation would respond to changes in humidity between 75 and 98%, whereas the pore size range corresponds to the change. It is assumed that the water stored in the mesopores will slowly release its water when the internal RH in concrete drops below 98% to compensate for the self-desiccation during hydration.

By using the concept of protected paste volume (Bentz and Snyder 1999), it revealed that cement paste should be closed to the internal curing water reservoir so that the absorbed water could be penetrated. Thus, cement paste is protected from self-desiccation by the absorbed water. To achieve this, the curing agent particle size should be as small as possible (Liu et al. 2017; Zhutovsky et al. 2011). Previous studies reported that smaller internal curing agent particles show better response in self curing action than those of larger one (Lura and Breugel 2000). Zhutovsky et al. (2004a, b) also revealed that decrease the internal curing particle sizes down to 4–2 mm would able to improve the efficiency of curing.

3.2.3 Superabsorbent Polymer (SAP)

SAP was initially developed during the 1980s and has since been widely used in forestry, agriculture, health supplies and in other fields given their potential as a water reservoir and their ability to expand and retain water (Buchholz and Graham 1998; Jensen and Hansen 2001a, b; Esteves et al. 2007; Siramanont et al. 2010;

3.2 Types of Self Curing Agents and Mechanism in Cementitious Materials

Mechtcherine and Reinhardt 2012; Dang et al. 2017). The capability of SAP has been used with cementitious materials in concrete to mitigate shrinkage (autogenous and drying) via self curing (Jensen and Hansen 2001a; Mechtcherine et al. 2009; Craeye et al. 2011; Snoeck et al. 2015; Wehbe and Ghahremaninezhad 2017; Tu et al. 2019) to enhance the durability toward freeze and thaw deterioration (Mönnig and Lura 2007; Mechtcherine and Reinhardt 2012; Wong 2017). SAPs are recognized as hydrogels, consisting of a three-dimensional cross-link network structure that can absorb a large volume of liquid compared to their mass because of osmotic pressure and expand to form an insoluble gel (Jensen and Hansen 2001a, b; Kim and Schlangen 2010; Ding et al. 2017; Wong 2017). A chemical reaction will eventuate when SAP is exposed to an aqueous solution, leading to shrinkage or swelling of the SAP. The absorption of SAP is driven by osmotic pressure, as illustrated in Fig. 3.3, before it develops the space between cross-links and polymer chains. The presence of osmotic pressure originates from a concentration gradient of moveable ions between the gel and solution (Mechtcherine and Reinhardt 2012; Wang et al. 2015; Snoeck et al. 2017; Wong 2017). The swollen SAPs then react as water reservoirs in the concrete. However, as the humidity in the concrete decreases, the absorbed water is pulled back into the cement paste capillary pores.

However, when applied SAP in a cement-based system to entrain water, it can substantially decrease the extension of absorption due to the high pH environment of the concrete mixture compared to freshwater (Liu et al. 2017). Nevertheless, the SAP able to absorb much more water than LWA to be used as a curing agent in self curing concrete. Meanwhile, the presence of osmotic pressure is from a concentration gradient of moveable ions between the gel and solution. This leads to the SAPs gradually releasing the absorbed water and leaving the voids (Lee et al. 2010, 2016; Wong 2017). Researchers from previous studies reported that the dosage (Jensen and Hansen 2002; Esteves 2009; Liu et al. 2017), type (Snoeck et al. 2014; Liu et al. 2017),

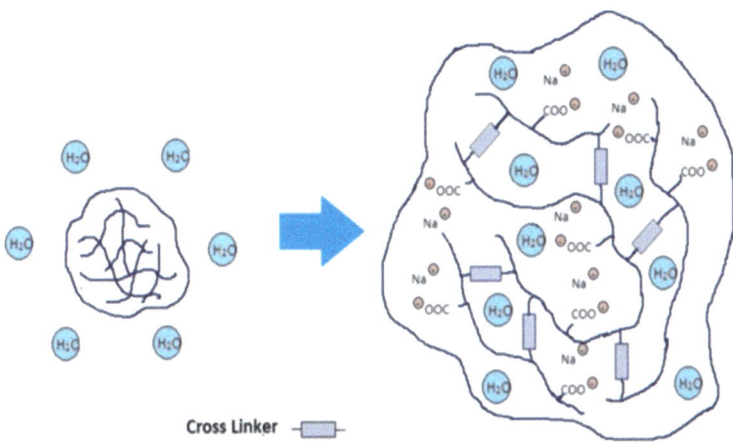

Fig. 3.3 Process of water uptake to SAP (Mechtcherine and Reinhardt 2012; licensed under a Creative Commons Attribution (CC BY))

particle size whereby SAP with bigger size able to absorb more water (Jensen and Hansen 2002; Esteves 2009) and water-saturated state of SAP (Igarashi and Watanabe 2006; Liu et al. 2017) acting as important role on effectiveness of self curing. SAPs were also found to improve autogenous healing (Kim and Schlangen 2010; Wong 2017) and tensile creep (Assmann and Reinhardt 2014; Shen et al. 2016; Wong 2017; Xin et al. 2018). Changing the rheology of the fresh material in concrete is another SAP characteristic (Mönnig 2009; Esteves 2010; Mechtcherine and Reinhardt 2012; Mechtcherine et al. 2015; Wong 2017). Dudziak and Mechtcherine (2010) and Schröfl (2012) reported that the workability of concrete containing SAP decreased due to extra internal curing water absorbed by the SAP.

3.2.4 Polyethylene Glycol (PEG)

Polyethylene-glycol is a condensation polymer of ethylene oxide and water with the general formula H $(OCH_2CH_2)_n$ OH, where n is the average number of repeating ox ethylene groups typically from 4 to about 180 (Dhir et al. 1994). According to Raoult's Law, when the vapor pressure of the solute in the pure condition is less than the vapor pressure of the solvent in the pure condition, it is apparent that theoretically, by adding additives, the vapor pressure of water will decrease, thus reducing the rate of evaporation process above the concrete surface (Dhir et al. 1994, 1998; Madduru et al. 2016; Sarbapalli et al. 2017; Sastry and Kumar 2018). Therefore, the application of water soluble polymers for instant polyethylene glycol (PEG) as self curing in concrete has been observed to be both effective and efficient in retaining water, reduces the water surface tension and enhancing the hydration process (El-Dieb 2007; El-Dieb et al. 2012; Mousa et al. 2015a, b; Sarbapalli et al. 2017; Madduru et al. 2018; El Wakkad et al. 2019; Sastry and Kumar 2018). Moreover, with water molecules, polymeric chains created by hydrophilic units form hydrogen bonds. A hydrogen bond is a frail bond formed in a compound between hydrogen atoms and strongly electronegative atoms in other molecules (Alberty and Daniels 1975; Dhir et al. 1994). The existence of a positive charge at the hydrogen atom causes attraction to the electronegative atom via electrostatically as illustrated in Fig. 3.4. To this end, water soluble polymers with either hydroxyl (–OH–) or ether (–O–) functional groups have been used as the chemical to minimize the impact of self-desiccation effect in concrete (Alberty and Daniels 1975; Dhir et al. 1994).

3.2.5 Natural Fibers

Previous researchers have investigated that wood-derived fibers and powder are potential as self curing agents in cement-based material due to the former capability to absorb and retain water in addition to gradually releasing absorbed water (Mohr 2005; Elsaid et al. 2011; Kawashima and Shah 2011; Mezencevova et al. 2011;

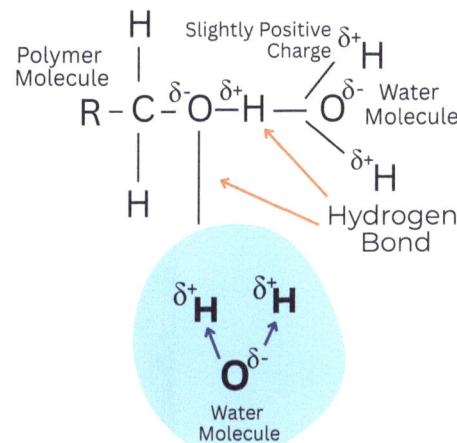

Fig. 3.4 Hydrogen bonds between water molecules and an –OH group on a polymer molecule (Dhir et al. 1994; Reproduced with permission from Springer Nature)

Zadeh and Bobko 2014; Jongvisuttisun et al. 2018). Good examples of wood-derived fibers used as self curing agent in concrete are eucalyptus pulp (Jongvisuttisun et al. 2018), kenaf fibers (Zadeh and Bobko 2014) and cellulose fibers (Kawashima and Shah 2011). Wood-derived fibers are hygroscopic materials and the movement of water via the pulps depends on the concentration gradient (diffusion) by capillary draw, and the effect of osmotic pressure (Wu et al. 2009; Mezencevova et al. 2011; Jongvisuttisun et al. 2018). Moavenzadeh (1990), Elsaid et al. (2011) and Jongvisuttisun et al. (2018) explained that wood-derived fibers consist of two pores namely larger pore (e.g. lumen) comprises of free water and smaller pore as illustrated in Fig. 3.5. Both pores play role in the transportation of moisture, from the wood pulp to nearby hydrating cement. Furthermore, the pore solution in cement-based material is an alkaline, which therefore, influences the character of wood pulp to swell or shrink and to change the effective size of the porous space (Lindstroem and Carlsson 1982; Scallan 1983; Bendzalova et al. 1996; Jongvisuttisun et al. 2018), thus affecting the water transport. Besides, free water in the lumen and weakly bound water may be released into the surrounding whereby the cement matrix may be self-desiccating over time. These would provide the matrix relief from self-desiccation and subsequently autogenous shrinkage will occur.

3.2.6 *Artificial Normal Weight Aggregate (ANWA)*

Waste material in the construction industry such as ceramic (Suzuki et al. 2009; Sato et al. 2011; Shigeta et al. 2018) and recycled concrete waste (Corinaldesi and Moriconi 2010; Silva et al. 2015; Jørgensen 2016; Suwan and Wattanachai 2017) have the potential as water reservoirs in self curing concrete given their ability to absorb water due to its porosity. Suzuki et al. (2009) and Meddah and Sato (2010) mentioned that saturated surface dry (SSD) density of crushed waste ceramic was

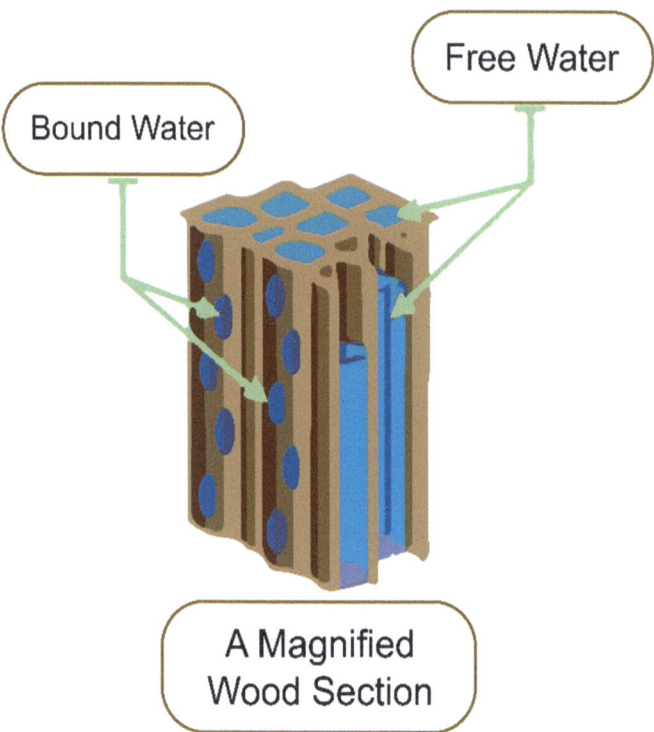

Fig. 3.5 Free and bound water in wood (adapted from Ahmed 2006)

2.48 g/cm^3, where the value is almost similar to that of natural normal fine aggregate. This statement is further strengthened by Shigeta et al. (2018), who reported that SSD density of waste ceramic coarse aggregate is 2.26 g/cm^3 while natural normal coarse aggregate is 2.62 g/cm^3. Thus, it provides a good effect on the strength of concrete containing crushed waste ceramic compared to that plain concrete. Moreover, water absorption of waste ceramic coarse aggregate was 9% compared to natural normal aggregate which recorded 0.67% water absorption. Meanwhile, Suzuki et al. (2009) revealed that water absorption of waste ceramic coarse aggregate also was 9% and the crushing rate value is 21.4% almost similar to that reported by Sato et al. (2011). The capability of water absorption in ceramic waste aggregates will help in the hydration process of the concrete.

3.3 Effects of Self Curing Agents on Properties of Concrete or Mortar

The following sub-sections present a review on workability, density, compressive, tensile and flexural strength on self cured concrete containing self curing agents.

3.3.1 Workability

Concrete has usually been coveted for its rapid strength development and good workability. However, it is frequently challenging to address both issues simultaneously when employing the traditional water-to-cement ratio. Increases in the water-to-cement ratio improve the workability of concrete but have the opposite effect on strength. Reducing the water-to-cement ratio of concrete to enhance its strength properties while keeping high workability is hard to achieve unless water-reducing admixtures are added to the mix. Several researchers have studied on the workability of mortar or concrete by using LWA as conventional aggregate replacement material. LWA is usually porous. The impact of porous aggregate on the workability of concrete is rarely discussed. The porous, water-absorbing lightweight particles can impair the workability of concrete and the effective water/binder ratio during the mixing stage (Al-Ani et al. 2020; Neville 2011; Nguyen et al. 2019). Water is often retained in LWA via holes that are bigger than those found in a hydrating cement paste (Dayalan, and Buellah 2007).

However, as the LWA content increased, the workability declined (Al-Ani et al. 2020; Chinnu et al. 2021). However, Khan et al. (2016) indicated that increasing oil palm shell (OPS) as aggregate replacement content increased the slump value up to a replacement level of 20%. Beyond that, the slump value decreased. The slump value decreased by 46 percent at a replacement level of 40% compared to control concrete due to the porous structure and higher water absorption of OPS in concrete than those made of granite aggregates. Additionally, the decline in the workability of concrete is mainly determined by the shape of the lightweight aggregate (Fanijo et al.2020; Okorafor et al. 2019).

On the other hand, the reduced workability in concrete can be effectively compensated by the use of pre-saturated of LWA. Soaked-porous aggregate neither absorbs nor releases water before setting. Hence, its workability is unaffected. If the LWA was used in saturated-surface dry (SSD) condition, part of the moisture near the surface of LWA may contribute actively to its workability. Therefore, the workability of concrete containing porous aggregate gave the same workability as conventional concrete. In this case, excess mixing water absorbed by porous aggregate does not affect the value of the effective water-cement ratio and does not reduce the strength of concrete (Zaichenko et al. 2015). Thus, a few minutes of pre-mixing of porous aggregate and water is also suggested (Castro et al. 2012; Ma et al. 2019). However, if the dry porous aggregate is added to the mixed concrete, the absorption rate will be slower, resulting in bleeding and segregation of the mixture at the first stage (Ma et al. 2019).

Studies have also shown that the addition of SAP caused workability reduction and delays the setting time of concrete (Dang et al. 2017; Piérard et al. 2006). Nevertheless, the pre-wetted SAP resulted in an increasing slump when SAP volume increased. It shows that the spherical particles that pre-absorbed SAP might act as lubricant in concrete mixture, reducing friction between paste and aggregate (Dang et al. 2017).

3.3.2 Density

Density is another important parameter to be tested for concrete containing LWA as self curing agent. The replacement of LWA for conventional coarse aggregates such as granite or limestone typically reduces concrete density. Shafigh et al. (2018) investigated the characteristics of fresh and hardened concrete utilizing lightweight expanded clay aggregate as a partial substitute for granite aggregate. When granite aggregates were replaced entirely with lightweight expanded clay aggregate, the density decreased from 2294 to 1507 kg/m^3. Due to the lower specific gravity of lightweight expanded clay aggregate than granite aggregates, the density of concrete reduced as lightweight expanded clay aggregate were added. A similar result was found by Shivashankar and Chetan (2018), which carried out for three different replacement levels. At replacement levels of 25% and 50%, the density was reduced by 24.4% and 48.8%, respectively, compared to the control concrete. A study conducted by Khan et al. (2016) revealed that an increase of oil coconut shells as lightweight aggregate in concrete decreased concrete density and compressive strength. This is because oil palm shell aggregates have a lower specific gravity and crushing strength than granite aggregates. After reaching a replacement level of 40%, a density reduction of approximately 16.5% was achieved compared to the reference concrete. Concrete weight reduction is beneficial in construction because it reduces the dead loads on structural elements.

3.3.3 Compressive Strength

Several researchers discovered that porous aggregate decreases the strength of high-performance concrete (HPC) and the strength stays decreased when the replacement of pre-wetted porous aggregate is increased (Zou and Weiss 2014; Raoufi et al. 2011; Costa 2012). Other researchers revealed that the reduction in concrete strength is due to the low strength of porous aggregate itself (Zhutovsky and Kovler 2017; Ma et al. 2019; Iffat et al. 2017; Hossain et al. 2012). The detrimental impact of porous aggregate on concrete strength can be mitigated by reducing the size of porous aggregate and improving its distribution (Liu et al. 2017; Gupta and Kua 2018). Generally, the results of compressive strength of concrete at earlier and later ages increased

3.3 Effects of Self Curing Agents on Properties of Concrete or Mortar

if the optimum proportion was obtained, usually about 20–40% replacement of conventional aggregate (Suwan and Wattanachai 2017). However, several researchers revealed that the compressive strength of concrete decreased if the replacement of porous aggregate was more than 50% (Akhnoukh 2017; Kim et al. 2016; Suwan and Wattanachai 2017; Chen et al. 2019) as shown in Fig. 3.6. The additional water supplied by the pre-wetted porous aggregate promotes a higher degree of hydration. It fills the pores with hydrated products (C–S–H gel), resulting in an improvement in compressive strength in concrete (Weber and Reinhardt 1997; Lura 2003). Agostini et al. (2010) discovered that porous aggregate reduced the amount of C–H while increasing the density of C–S–H at the interface between the aggregate and the cement paste.

Akhnoukh (2017) studied the influence of LWA addition to the compressive strength of normal strength concrete. The coarse LWA used is expanded shale, where the water absorption is 15% and specific gravity is 1.25. The replacement level used are 10%, 20% and 30%. Meanwhile, another series of specimens used lightweight expanded clay aggregates (LECA) was also prepared. The water absorption of LECA is 12.9% and specific gravity is 1.41. The replacement levels used are 12.5, 25, 37.5 and 50% of the total aggregate by volume. Compressive strength test was performed at age of 1, 7 and 28 days. It was concluded that the lower compressive strength was displayed by the concrete containing LECA. Table 3.1 shows the summary of percentage reduction of compressive strength for 28 days when LECA added in concrete as self curing agent. LWA can successfully replace up to 50% of normal-weight coarse aggregates without violating the minimum strength required.

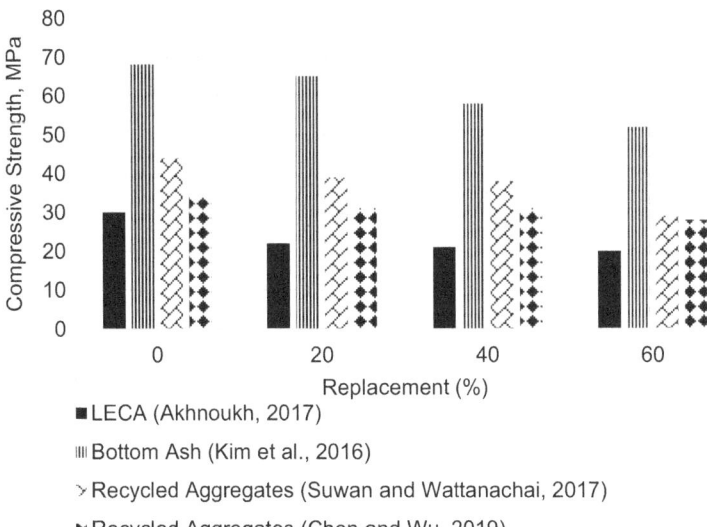

Fig. 3.6 Compressive strength of porous aggregate with percent replacement of conventional aggregate at 28 days

A similar result was found by Gopi and Revathi (2021). They used LECA and fly ash aggregates (FAA) as self curing agent in HPC. The fine aggregate was replaced by LECA/FAA under saturated surface dry (SSD) condition by 0–20% at a 5% interval in volume basis. The compressive strength was obtained at 7 and 28 days. It was observed that the compressive strength increased with an increase in LECA aggregate content from 0 to 15% replacement levels. The compressive strength decreases when LECA content exceeds 15%. The maximum compressive strength of 44.93 MPa was achieved by a mix containing 15% LECA (L_{15}) in the self curing concrete mix, as shown in Fig. 3.7.

Table 3.1 Compressive strength reduction of concrete due to LWA addition as compared to control for 28 days (Akhnoukh 2017)

Types of mixes	LWA percentage replacement (%)	Compressive strength
Expanded shale 1	10	8% < control
Expanded shale 2	20	9% < control
Expanded shale 3	30	18% < control
'Expanded clay 1	12.5	22% < control
Expanded clay 1	25	24% < control
Expanded clay 1	37.5	27% < control
Expanded clay 1	50	29% < control

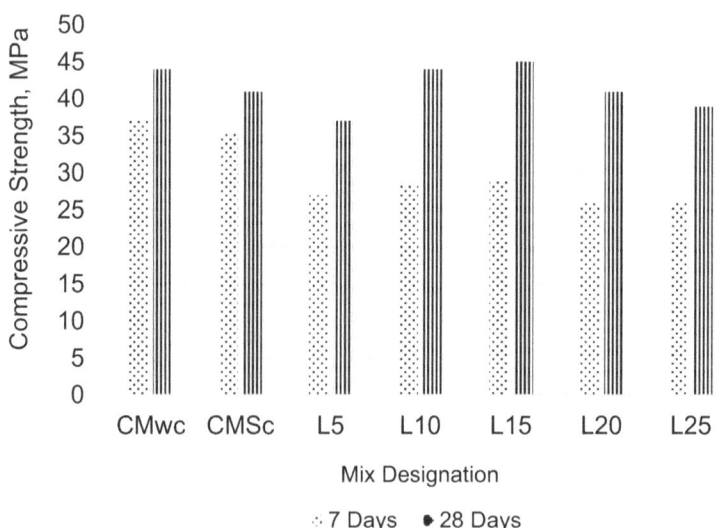

Fig. 3.7 Compressive strength for concrete containing different content of lightweight aggregate at 7 and 28 days (Reprinted from Gopi and Revathi 2021, Copyright 2021, with permission from Elsevier)

3.3 Effects of Self Curing Agents on Properties of Concrete or Mortar

It was discovered that LECA 15% is found to be optimum to develop self-compacting-curing concrete. Similar observations have been made by several researchers (Famili et al. 2012; Maghsoudi et al. 2011; Paul and Lopez 2011). Previous researchers reported many kinds of LWA to react as water reservoirs in concrete due to their water absorption ability, subsequently provide additional water for shrinkage mitigation, improve hydration and decrease the process of water evaporation in concrete (Liu et al. 2017; Zou et al. 2018). Expanded shale (Henkensiefken et al. 2010, 2011; Browning et al. 2011), expanded clay (Paul and Lopez 2011; Ghourchian et al. 2013; Zhang and Poon 2017; Pradeep and Beena 2020; Vázquez-Rodríguez et al. 2020), bottom ash (Kim et al. 2014; Wyrzykowski et al. 2016; Kim and Lee 2018; Balapour et al. 2020) and pumice (Zhutovsky et al. 2002a, b; Akcay and Tasdemir 2010; Zhutovsky and Kovler 2017; Vázquez-Rodríguez et al. 2020) are most common LWA used in concrete. However, existing LWAs as curing agents are highly porous in nature, resulting in more voids and decreased compressive strength in internally cured concrete (Hossain et al. 2012; Iffat et al. 2017; Zhutovsky and Kovler 2017; Ma et al. 2019; Kaplan et al. 2021).

A research study by Alaskar et al. (2021) on the effect of volcanic tuff as natural LWA on the compressive strength of HPC revealed that with the addition and increase in the natural LWA content, the compressive strength of concrete decreased. The compressive strength of 86 MPa was recorded at 28 days for the control mix without natural LWA. However, compressive strengths of 84, 77, and 70 MPa were obtained for mixes containing 5%, 10%, and 20% coarse natural LWA, respectively. The results indicated that the application of coarse natural LWA reduced the strength to 98, 89, and 81% of the control concrete strength, respectively, for replacement levels of 5, 10, and 20%.

Suwan and Wattanachai (2017) studied on recycled autoclaved aerated lightweight concrete (ACC- LWA) as aggregate. They suggested that 20–40% of LWA replacement can be an optimum for internal curing of high-strength concrete (HSC). The amount replacement of coarse (ACC- LWA) at percentage 20%, 40% and 60% respectively with the specific gravity 1.06, water absorption 30% and water cement ratio 0.35. Compressive strength test was carried out at age of 1, 3, 7 and 28 days. It was revealed that the compressive strength decreased with higher ACC-LWA proportion in both air and water curing as shown in Table 3.2.

Table 3.2 Compressive strength increment and decrement of concrete due to LWA addition as compared to control for 28 days (Suwan and Wattanachai 2017)

Types of mixes	Curing regimes	LWA percentage replacement (%)	Compressive strength
ACC-LWA 20	Water curing	20	9% < control
ACC-LWA 20	Air curing	20	6% > control
ACC-LWA 40	Water curing	40	14% < control
ACC-LWA 40	Air curing	40	3% < control
ACC-LWA 60	Water curing	60	37% < control
ACC-LWA 60	Air curing	60	29% < control

Paul and Lopez (2011), and Lura and Bisschop (2004) found that the use of an aggregate with a lower intrinsic strength than conventional aggregate could restrict the strength of the concrete achieved. Many investigations have been carried out to examine the effect of partial replacement of LWA as a self curing agent on the compressive strength of HPC. However, limited studies were published to examine the full replacement of conventional aggregate with LWA as a self curing agent on normal strength concrete.

Many researchers reported SAP improving the compressive strength (Jensen and Hansen 2001a, b, 2002; Ghourchian et al. 2013; Liu et al. 2017; Al Saffar et al. 2019; Memon et al. 2020) of HPC. However, some of the compressive strength decreased as greater amounts of SAP were added to the concrete (Woyciechowski and Kalinowski 2018; Song et al. 2016; Shen et al. 2016; Liu et al. 2021; Shen et al. 2020; Lei et al. 2020) as shown in Table 3.3. Song et al. (2016) studied the effect of SAP on the compressive strength of concrete and indicated that the decreasing of strength due to SAP augments in the concrete. SAP swelled after absorbing water, becoming hydrogels and acting as voids in the cementitious materials (Mechtcherine et al. 2013) as depicted in Fig. 3.8. The results also revealed that the ratio of early-age strength reduction due to SAP addition was more pronounced in the concrete specimen than in the concrete specimen without SAP. Nevertheless, the ratio of the later-age strength depended on the SAP dosage used in the specimen (Table 3.3). Strength development was increased due to the self curing by SAP, which resulted in enhanced hydration in the specimens. These results are comparable to observations in other studies (Hasholt et al. 2012; Snoeck et al. 2014).

The inclusion of water-soluble polymer self curing agent, polyethylene glycol (PEG) has significantly resulted in an increase of compressive strength in cementitious material compared to specimen without PEG, as reported by previous researchers (Rizzuto et al. 2020; Mousa et al. 2015a, b; Vaisakh et al. 2018). Mousa et al. (2015b) studied mixes with and without PEG, prepared and cured in laboratory air at 25 °C.

They found that the samples containing 2% of self curing agent showed 32.5% compressive strength increase at age of 28 days compared to the samples without self curing agent. The pronounced effect on the compressive strength might be due to normal concrete mixes that were air cured. Vaisakh et al. (2018) found that compressive strength air cured mixes with a PEG increase of 5.41% compared to water cured normal mixes. These reports are similar to the study by Rizzuto et al. (2020).

Meanwhile, a research study by Yew et al. (2020) indicated that a decrease was observed in compressive strength, flexural strength and split tensile strength as the proportion of light-expanded clay aggregates increased. The optimal outcomes were observed when conventional aggregates were substituted with expanded clay aggregates 70% of the time.

Several factors contribute to the decrease in compressive strength of LWA. As previously stated, the inclusion of LECA in concrete results in a reduction in density. Consequently, this leads to a corresponding decrease in compressive strength. It adheres to principle that the compressive strength of concrete increases with its density. Another factor is governed by the physical properties of LECA itself.

3.3 Effects of Self Curing Agents on Properties of Concrete or Mortar

Table 3.3 Compressive strength of mixtures with SAP at 28 days

Compressive strength, MPa

Researchers	w/c	Reference concrete	Cement-based with SAP addition (% by weight of binder)			
Lei et al. (2020)			0.1	0.2	0.3	0.8
	0.3	49	52	44	39	34
Shen et al. (2016)			0.05	0.16	0.26	–
	0.33	64	62.5	59.4	57.3	–
Woyciechowski and Kalinowski (2018)			0.05	0.09	0.14	–
	0.3	67	73	72	61	–
Liu et al. (2021)			0.2	0.4	0.6	–
	0.3	120	114	105	100	–
Shen et al. (2020)			0.57	0.86	1.14	–
	0.3	66.71	62.26	58.16	49.28	–
Piérard et al. (2006)			0.3	0.6	–	–
	0.3	107	99	93	–	–
Song et al. (2016)			0.15	0.3	–	–
	0.4	45	35	25	–	–

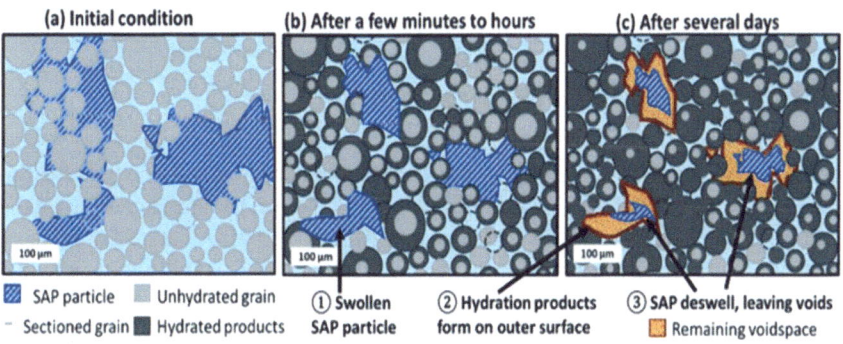

Fig. 3.8 a Soaked SAPs in the cement mixture; b SAPs gradually release the absorbed water; c SAPs de-swell and leave voids (Reprinted (adapted) with permission from Erk and Bose 2018. Copyright 2018 American Chemical Society)

Compared to normal-weight aggregates like granite, LECA has a higher void content, poorer particle strength, and crushing strength, which eventually weakens the strength of concrete. Abadel (2023) studied the compressive strength of Ultra High Performance Concrete (UHPC) with varying proportions of LWAs at the curing of 7, 28, and 56 days. The test findings indicated that when the proportion of LWAs increases from 0 to 30%, the compressive strength decreases at each corresponding curing

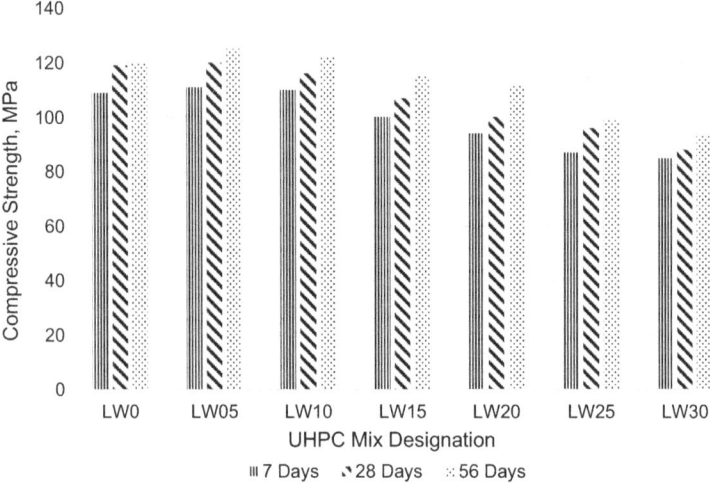

Fig. 3.9 Compressive strength of UHPC with varying proportions of LWAs: **a** at 7, 28, and 56 days of curing (Reproduced from Abadel 2023; licensed under a Creative Commons Attribution (CC BY))

period as shown in Fig. 3.9. At 56 days, the compressive strength increased by 4% and 1.31% with 5% and 10% LWA, respectively, compared to the sample without LWA. However, after this point, the compressive strength consistently decreased.

3.3.4 Split Tensile and Flexural Strength

Tensile and flexural strength are also parameters that used to determine the effect of internal curing. Several previous studies reported the effect of internal curing on the splitting tensile strength of HPC. Francis et al. (2017) reported that a 25% replacement of LWA results in the maximum strength (compressive, splitting tensile and flexural or modulus of rupture) for concrete grade M30. Flexural and splitting tensile strength has similar behavior as compressive strength, that decreased in flexural strength as the replacement of LWA increased (Hoff 2002). According to Kockal and Ozturan (2011), the minimum 28 days splitting tensile strength required for structural lightweight concrete to be used in structural element is 2.0 MPa. The flexural strength of normal weight concrete with a compressive strength of 34–55 MPa is in the range of 5–6 MPa with a flexural/compressive strength ratio of 11.6–13.5% (Mehta and Monteiro 2014). Zhutovsky and Kovler (2017) mentioned that internal curing at the water to cement ratios of 0.33, 0.25 and 0.21 had a detrimental influence on splitting tensile strength; this loss of splitting tensile strength is considerable at an early age. The results indicated that at the age of one day, the reduction percentages were 20%, 20% and 27% for water to cement ratios of 0.25, 0.21 and 0.33, respectively. On

3.3 Effects of Self Curing Agents on Properties of Concrete or Mortar

the other hand, the splitting tensile strength was comparable to that of the reference cured concrete.

According to Pradeep et al. (2019), which study on effect of pre-soaked LECA on strength, durability and flexural behavior of HPC found that the tensile strength of the mix with 15% LECA decreased by 2.4% at 28 days (Table 3.4). This could be owing to the porous characteristic of LECA, which allows a split plane to propagate through the aggregate. However, the flexural strength of the mix with 15% LECA showed an increment of 3.5% for the concrete specimen. The 15% replacement gave a maximum increase in strength of 17.5% at 28 days, and this percentage was taken as the optimum.

Despite the fact that the internal curing effect of LECA enhances strength, its brittleness might have a negative impact on the tensile properties of concrete. This may be the explanation for the inability of concrete to produce a higher tensile strength.

Mousa et al. (2015b) investigated the mechanical properties of self curing concrete found that the use of saturated LECA in the different ratios in concrete mixes allows internal curing for the concrete by providing continual hydration, which increases the tensile strength of concrete. The results indicated that the maximum improvement in tensile strength at 28 days was approximately 7.4% when saturated LECA was replaced with 15%, as depicted in Table 3.5. At an early age, concrete without LECA (ordinary concrete) displayed a higher flexural strength than concretes containing LECA. However, flexural strength rose gradually at 28 days, with increment percentages of 1.6%, 7.2%, and 3.4% for 10%, 15%, and 20% of LECA, respectively. The results of this experiment indicate that the tensile strength of all self curing concrete was between 6.4 and 8.5%.

Studies by Zou et al. (2018) on flexural strength of plain and internally cured mortar containing shale and coal bottom ash were revealed that the flexural strength of internally cured mortar is lower than that of the plain one. It is possible to conclude

Table 3.4 Flexural strength and split tensile strength for concrete specimen at 28 days (Pradeep et al. 2019)

Mix	Flexural strength		Split tensile strength	
	Strength	% increase	Strength	% decrease
Normal	6.20	–	4.385	–
15% LECA	6.42	3.5	4.279	2.4

Table 3.5 Effect of LECA on different mechanical properties of concrete at 28 days (Mousa et al. 2015b)

Type	Percent (%)	Tensile strength (%)	Flexural strength (%)
LECA	10	+ 3.7	+ 1.6
	15	+ 7.4	+ 7.2
	20	+ 5.6	+ 3.4

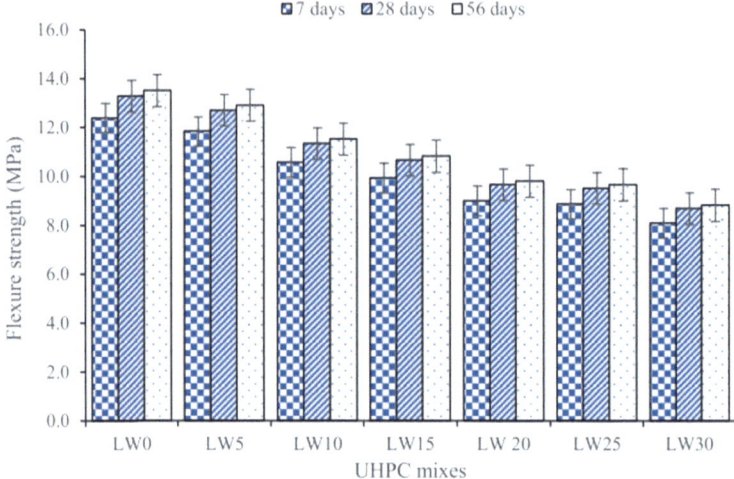

Fig. 3.10 Flexural strength of UHPC containing lightweight aggregate at 7, 28, and 56 days of curing (Abadel 2023; licensed under a Creative Commons Attribution (CC BY))

that aggregate strength has an intensity influence on flexural strength. The impact of internal curing is also distinct. The impact of internal curing on the strength of low water to cement ratio mortar resides in the fact that porous aggregate tends to reduce strength. Still, hydration enhances caused by internal curing benefits strength. Low porosity always equals higher strength for porous aggregate. The strength of a mortar accompanied by denser porous aggregate would be better. The impact of improved hydration is determined by the age and efficiency of porous aggregates. The later age and better efficiency of porous aggregate indicate that there is more internal curing water in a paste that could react with anhydrous cement particles. According to Abadel (2023), the flexural strength of UHPC samples with different amounts of lightweight aggregates after being cured for 7, 28, and 56 days shows that when the LWA percentage grows from 0 to 30%, there is a consistent decrease in flexural strength at each curing time as illustrated in Fig. 3.10. The decrease in flexural strength of the Ultra-High Performance Concrete (UHPC) with 0–30% lightweight pumice aggregates after 7, 28, and 56 days of curing may be due to many variables. Firstly, lightweight aggregates have poorer mechanical qualities compared to traditional aggregates (Li et al. 2018). Secondly, the reduced adhesion between the cement matrix and the lightweight aggregates can lead to a decrease in the binding strength at the interface between the aggregates and the cement paste (Zhuang et al. 2016). Secondly, in comparison to conventional aggregates, the lighter aggregates have a greater porosity and a reduced surface area, both of which may result in a weakened bond between the lightweight aggregates and the cement matrix. As a consequence, the flexural rigidity of the UHPC may be diminished.

References

S.Y. Abate, K. Song, S.J.K. Il, B.Y. Lee, H.K. Kim, Internal curing effect of raw and carbonated recycled aggregate on the properties of high-strength slag-cement mortar. Constr. Build. Mater. **165**, 64–71 (2018). https://doi.org/10.1016/j.conbuildmat.2018.01.03

A.A. Abadel, Physical, mechanical and microstructure characteristics of ultra-high-performance concrete containing lightweight aggregates. Materials **16**(13) (2023). https://doi.org/10.3390/ma16134883

F. Agostini, C. Davy, F. Skoczylas, T. Dubois, Effect of microstructure and curing conditions upon the performance of a mortar added with Treated Sediment Aggregates (TSA). Cem. Concr. Res. **40**, 1609–1619 (2010). https://doi.org/10.1016/j.cemcon-res.2010.07.003

Y. Agrawal, T. Gupta, R. Sharma, N.L. Panwar, S. Siddique, A comprehensive review on the performance of structural lightweight aggregate concrete for sustainable construction. Constr. Mater. **1**(1), 39–62 (2021). https://doi.org/10.3390/constrmater1010003

M.Z. Ahmed, Evaluation of moisture content in wood fiber and recommendation of the best method for its determination. Master Thesis (Helwan University, Cairo, 2006)

A.K. Akhnoukh, Internal curing of concrete using lightweight aggregates. Part. Sci. Technol. **36**, 362–367 (2017)

B. Akcay, M.A. Tasdemir, Effects of distribution of lightweight aggregates on internal curing of concrete. Cem. Concr. Compos. **32**(8), 611–616 (2010). https://doi.org/10.1016/j.cemconcomp.2010.07.003

I.A. Al-Ani, N. Wan Hamidon Al-Ansari, W.H. Mohtar, Development of lightweight concrete using industrial waste palm oil clinker. J. Civil Eng. Arch. **14**(6), 293–307 (2020). https://doi.org/10.17265/1934-7359/2020.06.001

A. Alaskar, M. Alshannag, M. Higaze, Mechanical properties and durability of high-performance concrete internally cured using lightweight aggregates. Constr. Build. Mater. **288**, 122998 (2021). https://doi.org/10.1016/j.conbuildmat.2021.122998

R.A. Alberty, F. Daniels, *Physical Chemistry,* 5th edn (Wiley, 1975) (May 30, 1979)

F.C.R. Almeida, A.J. Klemm, Efficiency of internal curing by superabsorbent polymers (SAP) in PC-GGBS mortars. Cem. Concr. Compos. **88**, 41–51 (2018). https://doi.org/10.1016/j.cemconcomp.2018.01.002

D.M. Al Saffar, A.J.K. Al Saad, B.A. Tayeh, Effect of internal curing on behavior of high performance concrete: an overview. Case Stud. Constr. Mater. **10**, e00229 (2019). https://doi.org/10.1016/j.cscm.2019.e00229

M.N. Amin, S. Hissan, K. Shahzada, K. Khan, T. Bibi, Pozzolanic reactivity and the influence of rice husk ash on early-age autogenous shrinkage of concrete. Front. Mater. **6**(July), 1–13 (2019). https://doi.org/10.3389/fmats.2019.00150

A. Assmann, H.W. Reinhardt, Tensile creep and shrinkage of SAP modified concrete. Cem. Concr. Res. **58**, 179–185 (2014). https://doi.org/10.1016/j.cemconres.2014.01.014

ASTM C1761, *Standard Specification for Lightweight Aggregate for Internal Curing of Concrete* (2017). https://doi.org/10.1520/C1761

M. Balapour, W. Zhao, E.J. Garboczi, N.Y. Oo, S. Spatar, Y.G. Hsuan et al., Potential use of lightweight aggregate (LWA) produced from bottom coal ash for internal curing of concrete systems. Cem. Concr. Compos. **105**, 103428 (2020). https://doi.org/10.1016/j.cemconcomp.2019.103428

M.M.H.W. Bandara, W.K. Mampearachchi, T. Anojan, Enhance the properties of concrete using pre-developed burnt clay chips as internally curing concrete aggregate. Case Stud. Constr. Mater. **11**, e00284 (2019). https://doi.org/10.1016/j.cscm.2019.e00284

H. Bari, M. Safiuddin, M.A. Salam, Microstructure of structural lightweight concrete incorporating coconut shell as a partial replacement of brick aggregate and its influence on compressive strength. Sustainability **13**(13), 7157 (2021). https://doi.org/10.3390/su13137157

A.A. Bashandy, Self-curing concrete under sulfate attack. Arch. Civil Eng. LXII **2**, 3–18 (2016)

M. Bendzalova, A. Pekarovicova, B. Kokta, R. Chen, Accessibility of swollen cellulosic fibers. Cellul. Chem. Technol. **30**(1), 19–32 (1996)

D.P. Bentz, K.A. Snyder, Protected paste volume in concrete: extension to internal curing using saturated lightweight fine aggregate. Cem. Concr. Res. **29**(11), 1863–1867 (1999). https://doi.org/10.1016/S0008-8846(99)00178-7

J. Browning, D. Darwin, D. Reynolds, B. Pendergrass, Lightweight aggregate as internal curing agent to limit concrete shrinkage. ACI Mater. J. **108**, 638–644 (2011)

F. Buchholz, A. Graham, *Modern superabsorbent polymer technology* (Wiley-VCH, 1998)

J. Castro, L. Keiser, M. Golias, J. Weiss, Absorption and desorption properties of fine lightweight aggregate for application to internally cured concrete mixtures. Cement Concr. Compos. **33**(10), 1001–1008 (2011). https://doi.org/10.1016/j.cemconcomp.2011.07.006

F. Chen, K. Wu, L. Ren, J. Xu, H. Zheng, Internal curing effect and compressive strength calculation of recycled clay brick aggregate concrete. Materials **12**, 1815 (2019). https://doi.org/10.3390/ma12111815

S.N. Chinnu, S.N. Minnu, A. Bahurudeen, R. Senthilkumar, Recycling of industrial and agricultural wastes as alternative coarse aggregates: a step towards cleaner production of concrete. Constr. Build. Mater. **287**, 123056 (2021). https://doi.org/10.1016/j.conbuildmat.2021.123056

V. Corinaldesi, G. Moriconi, Recycling of rubble from building demolition for low-shrinkage concretes. Waste Manage. **30**(4), 655–659 (2010). https://doi.org/10.1016/j.wasman.2009.11.026

H. Costa, E. Júlio, J. Lourenço, New approach for shrinkage prediction of high-strength lightweight aggregate concrete. Constr. Build. Mater. **35**, 84–91 (2012). https://doi.org/10.1016/j.conbuildmat.2012.02.052

B. Craeye, M. Geirnaert, G.D. Schutter, Super absorbing polymers as an internal curing agent for mitigation of early-age cracking of high-performance concrete bridge decks. Constr. Build. Mater. **25**(1), 1–13 (2011). https://doi.org/10.1016/j.conbuildmat.2010.06.063

J. Dang, J. Zhao, Z. Du, "Effect of superabsorbent polymer on the properties of concrete. Polymers **9**, 672 (2017). https://doi.org/10.3390/polym9120672

Dayalan J. and Buellah M., 2007. "Internal curing of concrete using prewetted light weight aggregates". *International Journal of Innovative Research in Science, Engineering and Technology (An ISO, 3297* (3): 2319–8753. www.ijirset.com.

R.K. Dhir, P.C. Hewlett, J. Lot, T.D. Dyer, An investigation into the feasibility of formulating "self-cure" concrete. Mater. Struct. **27**, 606–615 (1994)

R.K. Dhir, P.C. Hewlett, T.D. Dyer, Mechanisms of water retention in cement pastes containing a self-curing agent. Mag. Concr. Res. **50**(1), 85–90 (1998)

H. Ding, L. Zhang, P. Zhang, Factors influencing strength of super absorbent polymer (SAP) concrete. Trans. Tianjin Univ. **23**, 245–257 (2017). https://doi.org/10.1007/s12209-017-0049-y

L. Dudziak, V. Mechtcherine, Enhancing early-age resistance to cracking in high-strength cement based materials by means of internal curing using super absorbent polymers. Int. RILEM Conf. Mater. Sci. **III**, 129–139 (2010)

A.S. El-Dieb, Self-curing concrete: Water retention, hydration and moisture transport. Constr. Build. Mater. **21**, 1282–1287 (2007). https://doi.org/10.1016/j.conbuildmat.2006.02.007

A.S. El-Dieb, T.A. El-Maaddawy, A.A.M. Mahmoud, Water-soluble polymers as selfcuring agents in cement mixes. Adv. Cem. Res. **24**(5), 291–299 (2012). https://doi.org/10.1680/adcr.11.00030

A. Elsaid, M. Dawood, R. Seracino, C. Bobko, Mechanical properties of kenaf fiber reinforced concrete. Constr. Build. Mater. **25**(4), 1991–2001 (2011). https://doi.org/10.1016/j.conbuildmat.2010.11.052

N. El Wakkad, K.M. Heiza, A. Eladly, in *11th International Conference on Nano Technology in Construction*. Review on Self-curing Concrete (2019), pp. 1–16

K.A. Erk, B. Bose, *Using Polymer Science to Improve Concrete: Superabsorbent Polymer Hydrogels in Highly Alkaline Environments* (American Chemical Society (ACS), Washington, DC, 2018), pp. 333–356

References

L.P. Esteves, P. Cachim, V.M. Ferreira, in *Advances in Construction Materials*. Mechanical Properties of Cement Mortars with Superabsorbent Polymers (2007), pp. 451–462. https://doi.org/10.1007/978-3-540-72448-3

P. Esteves Luís, *Internal Curing in Cement-Based Materials*. Ph.D. Thesis (Universidade de Aveiro Aveiro, Portugal, 2009)

L.P. Esteves, in *International RILEM Conference on Use of Superabsorbent Polymers and Other New Additives in Concrete*. On the Absorption Kinetics of Superabsorbent Polymers (2010), pp. 77–84

H. Famili, S.M. Khodadad, T. Parhizkar, Internal curing of high strength self consolidating concrete by saturated lightweight aggregate—effects on material properties. Int. J. Civil Eng. **10**(3), 210–221 (2012)

E. Fanijo, A.J. Babafemi, O. Arowojolu, Performance of laterized concrete made with palm kernel shell as replacement for coarse aggregate. Constr. Build. Mater. **250**, 118829 (2020). https://doi.org/10.1016/j.conbuildmat.2020.118829

S.J. Francis, B. Karthik, H. Gokulram, Flexural behaviour of self-curing concrete with lightweight aggregate and polyethylene glycol. Int. J. Eng. Trends Technol. **47**(2), 71–77 (2017). https://doi.org/10.14445/22315381/ijett-v47p211

S. Ghourchian, M. Wyrzykowski, P. Lura, M. Shekarchizadeh, B. Ahmadi, An investigation on the use of zeolite aggregates for internal curing of concrete. Constr. Build. Mater. **40**, 135–144 (2013). https://doi.org/10.1016/j.conbuildmat.2012.10.009

R. Gopi, V. Revathi, Flexural behavior of self compacting self curing concrete with lightweight aggregates. Mater. Today Proc. **45**, 2449–2455 (2021). https://doi.org/10.1016/j.matpr.2020.11.019

Z.C. Grasley, D.A. Lange, M.D. D'Ambrosia, Internal relative humidity and drying stress gradients in concrete. Mater. Struct./materiaux Et Constr. **39**(9), 901–909 (2006). https://doi.org/10.1617/s11527-006-9090-3

S. Gupta, H.W. Kua, Effect of water entrainment by pre-soaked biochar particles on strength and permeability of cement mortar. Constr. Build. Mater. **159**, 107–125 (2018). https://doi.org/10.1016/j.conbuildmat.2017.10.095

M.T. Hasholt, O.M. Jensen, K. Kovler, S. Zhutovsky, Can superabsorent polymers mitigate autogenous shrinkage of internally cured concrete without compromising the strength? Constr. Build. Mater. **31**, 226–230 (2012). https://doi.org/10.1016/j.conbuildmat.2011.12.062

R. Henkensiefken, B. Peter, B. Dale, N. Tommy, W. Jason, Plastic shrinkage cracking in internally cured mixtures made with pre-wetted lightweight aggregate. Concr. Int. **32**(2), 49–64 (2010)

R. Henkensiefken, N. Tommy, W. Jason, Saturated lightweight aggregate for internal curing in low w/c mixtures: Monitoring water movement using x-ray absorption. Int. J. Exp. Mech. **47**, 432–41 (2011). https://doi.org/10.1111/j.1475-1305.2009.00626.x

G.C. Hoff, The use of lightweight fines for the internal curing of concrete. Report Northeast Solite Corp. 1–44 (2002)

T. Hossain, A. Salam, M.A. Kader, Pervious concrete using brick chips as coarse aggregate: an experimental study. J. Civil Eng. **40**, 125–137 (2012)

S. Iffat, T. Manzur, M.A. Noor, Durability performance of internally cured concrete using locally available low cost LWA. KSCE J. Civil Eng. **21**, 1256–1263 (2017). https://doi.org/10.1007/s12205-016-0793-x

S. Igarashi, A. Watanabe, in *Proceedings of the International RILEM Conference*. Experimental Study on Prevention of Autogenous Deformation by Internal Curing Using Super-Absorbent Polymer Particles (2006), pp. 77–86

O.M. Jensen, P.F. Hansen, Autogenous deformation and RH-change in perspective. Cem. Concr. Res. **31**(12), 1859–1865 (2001). https://doi.org/10.1016/S0008-8846(01)00501-4

O.M. Jensen, P.F. Hansen, Water-entrained cement-based materials I. Principles and theoretical background. Cem. Concr. Res. **31**(4), 647–654 (2001)

O.M. Jensen, P.F. Hansen, Water-entrained cement-based materials II: experimental observations. Cem. Concr. Res. **32**(6), 973–978 (2002). https://doi.org/10.1016/S0008-8846(02)00737-8

P. Jongvisuttisun, J. Leisen, K.E. Kurtis, Key mechanisms controlling internal curing performance of natural fibers. Cem. Concr. Res. **107**, 206–220 (2018). https://doi.org/10.1016/j.cemconres.2018.02.007

L. Jørgensen, *Concrete Waste as Internal Water Reservoir in High Strength Concrete* (2016)

M.M. Kamal, M.A. Safan, A.A. Bashandy, A.M. Khalil, Experimental investigation on the behavior of normal strength and high strength self-curing self-compacting concrete. J. Build. Eng. **16**, 79–93 (2018). https://doi.org/10.1016/j.jobe.2017.12.012

S. Kang, S. Hong, J. Moon, Importance of drying to control internal curing effects on field casting ultra-high performance concrete. Cem. Concr. Res. **108**, 20–30 (2018). https://doi.org/10.1016/j.cemconres.2018.03.008

G. Kaplan, G. Aslinur, C. Betul, Y.B. Oguzhan, The impact of recycled coarse aggregates obtained from waste concretes on lightweight pervious concrete properties. Environ. Sci. Pollut. Res. **28**(14), 17369–94 (2021). https://doi.org/10.1007/s11356-020-11881-y

S. Kawashima, S.P. Shah, Early-age autogenous and drying shrinkage behavior of cellulose fiber-reinforced cementitious materials. Cement Concr. Compos. **33**(2), 201–208 (2011). https://doi.org/10.1016/j.cemconcomp.2010.10.018

J.T. Kevern, Q.C. Nowasell, Internal curing of pervious concrete using lightweight aggregates. Constr. Build. Mater. **161**, 229–235 (2018). https://doi.org/10.1016/j.conbuildmat.2017.11.055

M.M.H. Khan, W.L. Guong, T.J. Deepak, S. Nai, Use of oil palm shell as replacement of coarse aggregate for investigating properties of concrete. Int. J. Appl. Eng. Res. **11**(4), 2379–2383 (2016)

J.S. Kim, E. Schlangen, in *2nd International Symposium on Service Life Design for Infrastructure*. Super Absorbent Polymers to Stimulate Self Healing in ECC (2010), pp. 849–858

H.K. Kim, J. Jang, Y.C. Choi, H. Lee, Improved chloride resistance of high-strength concrete amended with coal bottom ash for internal curing. Constr. Build. Mater. **71**, 334–343 (2014). https://doi.org/10.1016/j.conbuildmat.2014.08.069

H.K. Kim, K. Ha, H. Lee, Internal-curing efficiency of cold-bonded coal bottom ash aggregate for high-strength mortar. Constr. Build. Mater. **126**, 1–8 (2016). https://doi.org/10.1016/j.conbuildmat.2016.08.125

H.K. Kim, H. Lee, Hydration kinetics of high-strength concrete with untreated coal bottom ash for internal curing. Cem. Concre. Compos. **91**, 67–75 (2018). https://doi.org/10.1016/j.cemconcomp.2018.04.017

N.U. Kockal, T. Oztura, Strength and elastic properties of structural lightweight concretes. Mater. Des. **32**(4), 2396–2403 (2011). https://doi.org/10.1016/j.matdes.2010.12.053

N. Kon, S. Yihune, H. Kim, Use of recycled aggregates as internal curing agent for alkali-activated slag system. Constr. Build. Materi. **159**, 286–296 (2018). https://doi.org/10.1016/j.conbuildmat.2017.10.110

H.X.D. Lee, H.S. Wong, N.R. Buenfeld, Potential of superabsorbent polymer for self-sealing cracks in concrete. Adv. Appl. Ceram. **109**(5), 296–302 (2010). https://doi.org/10.1179/174367609X459559

X. Lei, R. Wang, H. Jiang, F. Xie, Y. Bao, Effect of internal curing with superabsorbent polymers on bond behavior of high-strength concrete. Adv. Mater. Sci. Eng. 1–13 (2020). https://doi.org/10.1155/2020/6651452

Y. Li, K.H. Tan, E.H. Yang, Influence of aggregate size and inclusion of polypropylene and steel fibers on the hot permeability of ultra-high performance concrete (UHPC) at elevated temperature. Constr. Build. Mater. **169**, 629–637 (2018). https://doi.org/10.1016/j.conbuildmat.2018.01.105

T. Lindstroem, G. Carlsson, The effect of carboxyl groups and their ionic form during drying on the hornification of cellulose fibers. Svensk Papperstidning **85**(15), 146–151 (1982)

J. Liu, C. Shi, X. Ma, K. Khayat, J. Zhang, D. Wang, An overview on the effect of internal curing on shrinkage of high performance cement-based materials. Constr. Build. Mater. **146**, 702–712 (2017). https://doi.org/10.1016/j.conbuildmat.2017.04.154

References

J. Liu, C. Shi, N. Farzadnia, X. Ma, Effects of pretreated fine lightweight aggregate on shrinkage and pore structure of ultra-high strength concrete. Constr. Build. Mater. **204**, 276–287 (2019). https://doi.org/10.1016/j.conbuildmat.2019.01.205

J. Liu, N. Farzadnia, K.H. Khayat, C. Shi, Effects of SAP characteristics on internal curing of UHPC matrix. Constr. Build. Mater. **280** (2021) https://doi.org/10.1016/j.conbuildmat.2021.122530

P. Lura, K.V. Breugel, in *International RILEM Workshop on Shrinkage of Concrete*. Moisture Exchange as a Basic Phenomenon to Understand Volume Changes of Lightweight Aggregate Concrete at Early Age (2000)

P. Lura, *Autogenous Deformation and Internal Curing of Concrete* (IOS Press, Amsterdam, 2003), pp.1–180

P. Lura, J. Bisschop, On the origin of eigenstresses in lightweight aggregate concrete. Cem. Concr. Compos. **26**, 445–452 (2004). https://doi.org/10.1016/s0958-9465(03)00072-6

P. Lura, M. Wyrzykowski, C. Tang, E. Lehmann, Internal curing with lightweight aggregate produced from biomass-derived waste. Cem. Concr. Res. **59**, 24–33 (2014). https://doi.org/10.1016/j.cemconres.2014.01.025

X. Ma, J. Liu, C. Shi, A review on the use of LWA as an internal curing agent of high performance cement-based materials. Constr. Build. Mater. **218**, 385–393 (2019). https://doi.org/10.1016/j.conbuildmat.2019.05.126

S.R.C. Madduru, S.N.R.G. Pallapothu, R.K. Pancharathi, R.K. Garje, R. Chakilam, Effect of self curing chemicals in self compacting mortars. Constr. Build. Mater. **107**, 356–364 (2016). https://doi.org/10.1016/j.conbuildmat.2016.01.018

S.R.C. Madduru, P.R. Kumar, P.S.N.R. Giri, G.R. Kumar, Performance and microstructure characteristics of self curing self compacting concrete. Adv. Cem. Res. **30**(10), 451–468 (2018). https://doi.org/10.1680/jadcr.17.00154

A.A. Maghsoudi, S. Mohamadpour, M. Maghsoudi, Mix design and mechanical properties of self compacting light weight concrete. Int. J. Civil Eng. **9**(3), 230–236 (2011)

T. Manzur, S. Rahman, T. Torsha, M.A. Noor, K.M.A. Hossain, Burnt clay brick aggregate for internal curing of concrete under adverse curing conditions. KSCE J. Civil Eng. **23**(12), 5143–5153 (2019). https://doi.org/10.1007/s12205-019-0834-3

V. Mechtcherine, L. Dudziak, S. Hempel, in *Proceedings of the 8th International Conference on Creep, Shrinkage and Durability Mechanics of Concrete and Concrete Structures*. Mitigating Early Age Shrinkage of Ultra-High Performance Concrete by Using Super Absorbent Polymers (SAP). (2009), pp. 847–853

V. Mechtcherine, H.W. Reinhardt, *Application of Superabsorbent Polymers (SAP) in Concrete Construction* (Springer, 2012)

V. Mechtcherine, M. Gorges, C. Schroefl, A. Assmann, W. Brameshuber, A.B. Ribeiro, D. Cusson, J. Custódio, E.F. da Silva, K. Ichimiya, "Effect of internal curing by using superabsorbent polymers (SAP) on autogenous shrinkage and other properties of a high-performance fine-grained concrete". Results of a RILEM round-robin test. Mater. Struct. **47**, 541–562 (2013). https://doi.org/10.1617/s11527-013-0078-5

V. Mechtcherine, E. Secrieru, C. Schröfl, Effect of superabsorbent polymers (SAPs) on rheological properties of fresh cement-based mortars—development of yield stress and plastic viscosity over time. Cem. Concr. Res. **67**, 52–65 (2015). https://doi.org/10.1016/j.cemconres.2014.07.003

M.S. Meddah, R. Sato, Effect of curing methods on autogenous shrinkage and self-induced stress of high-performance concrete. ACI Mater. J. **107**(1), 65–74 (2010). https://doi.org/10.14359/51663467

P.K. Mehta, P.J.M. Monteiro, *Concrete: Microstructure, Properties, and Materials*, 4th edn (2014)

R.P. Memon, A.R.M. Sam, A.Z. Awang, M.M. Tahir, A. Mohamed, K.A. Kassim, A. Ismail, Introducing effective microorganism as self-curing agent in self-cured concrete. IOP Conf. Ser. Mater. Sci. Eng. **849**, 012081 (2020)

A. Mezencevova, V. Garas, H. Nanko, K.E. Kurtis, Influence of thermomechanical pulp fiber compositions on internal curing of cementitious materials. J. Mater. Civ. Eng. **24**(8), 970–975 (2011). https://doi.org/10.1061/(asce)mt.1943-5533.0000446

F. Moavenzadeh, *Concise Encyclopedia of Building and Construction Materials.* Pergamon Press (1990). https://doi.org/10.1520/jte12552j

D. Mohamad, S. Beddu, S.N. Sadon, N.L. Mohd Kamal, Z. Itam, M.A. Zainol, M.Z. Ramli, W.M. Sapuan, Properties of self-curing concrete containing bottom ash. Int. J. Adv. Appl. Sci. **4**(11), 138–142 (2017)

B.J. Mohr, *Durability of Pulp Fiber-Cement Composites.* PhD Dissertation (Georgia Institute of Technology, 2005). https://doi.org/10.1017/CBO9781107415324.004

S. Mönnig, P. Lura, Superabsorbent polymers—an additive to increase the freeze-thaw resistance of high strength concrete. Conf. Adv. Constr. Mater. 351–358 (2007). https://doi.org/10.1007/978-3-540-72448-3_35

M.I. Mousa, M.G. Mahdy, A.H. Abdel-reheem, A.Z. Yehia, Self-curing concrete types; water retention and durability. Alex. Eng. J. **54**(3), 565–575 (2015). https://doi.org/10.1016/j.aej.2015.03.027

M.I. Mousa, M.G. Mahdy, A.H. Abdel-Reheem, A.Z. Yehia, Mechanical properties of self-curing concrete (SCUC). HBRC J. **11**(3), 311–320 (2015). https://doi.org/10.1016/j.hbrcj.2014.06.00

R. Mrad, G. Chehab, Mechanical and microstructure properties of biochar-based mortar: an internal curing agent for PCC. Sustainability **11**(9) (2019). https://doi.org/10.3390/su11092491

Ş Mustafa, M. Lachemi, K.M.A. Hossain, V.C. Li, Internal curing of engineered cementitious composites for prevention of early age autogenous shrinkage cracking. Cem. Concr. Res. **39**, 893–901 (2009). https://doi.org/10.1016/j.cemconres.2009.07.006

A.M. Neville, *Properties of Concrete.* Pearson Education Limited (2011)

H.D. Nguyen, H.Q. Le, Water movement in internally cured concrete. IOP Conf. Ser. Mater. Sci. Eng. 1–8 (2018). https://doi.org/10.1088/1757-899X/365/3/032029

T.B.T. Nguyen, W. Saengsoy, S. Tangtermsirikul, "Influence of bottom ashes with different water retainabilities on properties of expansive mortars and expansive concretes. Eng. J. Thailand **23**(5), 107–123 (2019). https://doi.org/10.4186/ej.2019.23.5.107

S. Oh, Y. Cheol, Superabsorbent polymers as internal curing agents in alkali activated slag mortars. Constr. Build. Mater. **159**, 1–8 (2018). https://doi.org/10.1016/j.conbuildmat.2017.10.121

O.O. Okorafor, C.C. Egwuonwu, E.U. Ujah, M.I. Chikwue, N.G.J. Nadieze, Utilization of palm kernel shell (PKS) as coarse aggregate in lightweight concrete (LWC). Int. Res. J. Eng. Technol. **6**(10), 1730–1743 (2019)

Á. Paul, M. Lopez, Assessing lightweight aggregate efficiency for maximizing internal curing performance. ACI Mater. J. **108**(4), 385–393 (2011)

J. Piérard, V. Pollet, N. Cauberg, in *Proceedings of the International RILEM Conference Volume Changes Hardening Concrete: Testing Mitigation, Lyngby, Denmark, 20–23 August 2006.* Mitigating autogenous shrinkage in HPC by internal curing using superabsorbent polymers (2006), pp. 97–106. https://doi.org/10.1617/2351580052.011

P. Pradeep, M.O.L. Beena, Comparative study on the effect of lightweight aggregates on the properties of high performance concrete. J. Eng. Sci. Technol. **15**(1), 305–319 (2020)

P. Pradeep, Beenamol, H.S. Nair, Effect of pre-soaked light expanded clay aggregate on strength, durability and flexural behaviour of high-performance concrete. J. Eng. Sci. Technol. **14**(5), 2629–2642 (2019)

B.F. Rahmasari, Y. Yu, J. Yang, in *Sustainable Buildings and Structures: Building a Sustainable Tomorrow.* Overview of the Influence of Internal Curing in Recycled Aggregate Concrete (2019), p. 416

K. Raoufi, J. Schlitter, D. Bentz, W. Weiss, Parametric assessment of stress development and cracking in internally cured restrained mortars experiencing autogenous deformations and thermal loading. Adv. Civil Eng. (ID 870128), 1–16 (2011). https://doi.org/10.1155/2011/870128

S. Riyazi, J.T. Kevern, M. Mulheron, Super absorbent polymers (SAPs) as physical air entrainment in cement mortars. Constr. Build. Mater. **147**, 669–676 (2017). https://doi.org/10.1016/j.conbuildmat.2017.05.001

References

J.P. Rizzuto, K. Mounir, E. Hanaa, A. Bashandy, E. Zeinab, N. Mohamed, R. Aboel, G.S. Ibrahim, Effect of Self-curing admixture on concrete properties in hot climate conditions. Constr. Build. Mater. **261**, 119933 (2020). https://doi.org/10.1016/j.conbuildmat.2020.119933

D. Sarbapalli, Y. Dhabalia, K. Sarkar, B. Bhattacharjee, Application of SAP and PEG as curing agents for ordinary cement-based systems: impact on the early age properties of paste and mortar with water-to-cement ratio of 0.4 and above. Eur. J. Environ. Civ. Eng. **21**(10), 1237–1252 (2017). https://doi.org/10.1080/19648189.2016.1160843

G.K. Sastry, P.M. Kumar, Self-curing concrete with different self-curing agents. IOP Conf. Ser. Mater. Sci. Eng. **330**, 1–7 (2018). https://doi.org/10.1088/1757-899X/330/1/012120

R. Sato, A. Shigematsu, T. Nukushina, M. Kimura, Improvement of properties of portland blast furnace cement Type B concrete by internal curing using ceramic roof material waste. J. Mater. Civ. Eng. **23**, 777–782 (2011). https://doi.org/10.1061/(ASCE)MT.1943-5533.0000232

A.M. Scallan, The effect of acidic groups on the swelling of pulps: a review. Tappi J. **66**, 73–75 (1983)

C. Schröfl, V. Mechtcherine, M. Gorges, Relation between the molecular structure and the efficiency of superabsorbent polymers (SAP) as concrete admixture to mitigate autogenous shrinkage. Cem. Concr. Res. **42**(6), 865–873 (2012). https://doi.org/10.1016/j.cemconres.2012.03.011

G.R. de Sensale, B.R. António, A. Gonçalves, Effects of RHA on autogenous shrinkage of Portland cement pastes. Cem. Concr. Compos. **30**(10), 892–97 (2008). https://doi.org/10.1016/j.cemconcomp.2008.06.014

G.R. de Sensale, A.F. Goncalves, Effects of fine LWA and SAP as internal water curing agents. Int. J. Conr. Struct. Mater. **8**(3), 229–238 (2014). https://doi.org/10.1007/s40069-014-0076-1

P. Shafigh, L.J. Chai, H.B. Mahmud, M.A. Nomeli, A comparison study of the fresh and hardened properties of normal weight and lightweight aggregate concretes. J. Build. Eng. **15**(November 2017), 252–260 (2018). https://doi.org/10.1016/j.jobe.2017.11.025

D. Shen, X. Wang, D. Cheng, J. Zhang, G. Jiang, Effect of internal curing with super absorbent polymers on autogenous shrinkage of concrete at early age. Constr. Build. Mater. **106**(1), 512–522 (2016). https://doi.org/10.1016/j.conbuildmat.2015.12.115

D. Shen, C. Liu, J. Jiang, J. Kang, M. Li, Influence of super absorbent polymers on early-age behavior and tensile creep of internal curing high strength concrete. Constr. Build. Mater. **258**, 120068 (2020). https://doi.org/10.1016/j.conbuildmat.2020.120068

A. Shigeta, Y. Ogawa, K. Kawai, in *The 4th International Conference on Rehabilitation and Maintenance in Civil Engineering (ICRMCE 2018)*, vol. 195. Microscopic Investigation on Concrete Cured Internally by Using Porous Ceramic Roof-Tile Waste Aggregate (2018), pp. 1–7

B.C. Shivashankar, Chetan, An experimental investigation of light weight concrete by partial replacement of coarse aggregate as LECA. Int. J. Sci. Technol. Eng. **5**(1), 47–53 (2018)

J. Siramanont, W. Vichit-Vadakan, W. Siriwatwechakul, in *International RILEM Conference on Use of Superabsorbent Polymers and Other New Additives in Concrete*. The Impact of SAP Structure on the Effectiveness of Internal Curing (2010), pp. 1–10

R.V. Silva, J. De Brito, R.K. Dhir, Prediction of the shrinkage behavior of recycled aggregate concrete: a review. Constr. Build. Mater. **77**, 327–339 (2015). https://doi.org/10.1016/j.conbuildmat.2014.12.102

D. Snoeck, D. Schaubroeck, P. Dubruel, N. de Belie, Effect of high amounts of superabsorbent polymers and additional water on the workability, microstructure and strength of mortars with a water-to-cement ratio of 0.50. Constr. Build. Mater. **72**, 148–157 (2014). https://doi.org/10.1016/j.conbuildmat.2014.09.012

D. Snoeck, O.M. Jensen, N. De Belie, The influence of superabsorbent polymers on the autogenous shrinkage properties of cement pastes with supplementary cementitious materials. Cem. Concr. Res. **7**, 59–67 (2015). https://doi.org/10.1016/j.cemconres.2015.03.020

D. Snoeck, L. Pel, N. De Belie, The water kinetics of superabsorbent polymers during cement hydration and internal curing. Sci. Rep. (August), 1–14 (2017). https://doi.org/10.1038/s41598-017-10306-0

C. Song, Y.C. Choi, S. Choi, Effect of internal curing by superabsorbent polymers—internal relative humidity and autogenous shrinkage of alkali-activated slag mortars. Constr. Build. Mater. **123**, 198–206 (2016). https://doi.org/10.1016/j.conbuildmat.2016.07.007

T. Suwan, P. Wattanachai, Properties and internal curing of concrete containing recycled autoclaved aerated lightweight concrete as aggregate. Adv. Mater. Sci. Eng. 1–11 (2017). https://doi.org/10.1155/2017/2394641

M. Suzuki, M. Seddik, R. Sat, Use of porous ceramic waste aggregates for internal curing of high-performance concrete. Cem. Concr. Res. **39**(5), 373–381 (2009). https://doi.org/10.1016/j.cemconres.2009.01.007

W. Tu, Y. Zhu, G. Fang, X. Wan, M. Zhang, Internal curing of alkali-activated fly ash-slag pastes using superabsorbent polymer. Cem. Concr. Res. **116**, 179–190 (2019). https://doi.org/10.1016/j.cemconres.2018.11.018

N.V. Tuan, G. Ye, K.V. Breugel, O. Copuroglu, Hydration and microstructure of ultra high performance concrete incorporating rice husk ash. Cem. Concr. Res. **41**(11), 1104–1111 (2011). https://doi.org/10.1016/j.cemconres.2011.06.009

V. Van, C. Rößler, D. Bui, H. Ludwig, Rice husk ash as both pozzolanic admixture and internal curing agent in ultra-high performance concrete. Cem. Concr. Compos. **53**, 270–278 (2014). https://doi.org/10.1016/j.cemconcomp.2014.07.015

G. Vaisakh, M.S.R. Kumar, P.S. Bala, An experimental study on properties of M50 concrete cured using PEG 400. Int. J. Civil Eng. Technol. **9**, 725–732 (2018)

F.J. Vázquez-Rodríguez, N. Elizondo-Villareal, L.H. Verástegui, A.M. Arato Tovar, J.F. López-Perales, J.E. Contreras de León, C. Gómez-Rodríguez, D. Fernández-González, L.F. Verdeja, L.V. García-Quiñonez, E.A. Rodríguez Castellanos, Effect of mineral aggregates and chemical admixtures as internal curing agents on the mechanical properties and durability of high-performance concrete. Materials **13**(9), 2090 (2020). https://doi.org/10.3390/ma13092090

F. Wang, J. Yang, H. Cheng, J. Wu, X. Liang, Study on mechanism of desorption behavior of saturated superabsorbent polymers in concrete. ACI Mater. J. **112**(3), 463–469 (2015). https://doi.org/10.14359/51686996

J. Wang, F. Liu, in *International Concrete Sustainability Conference*. Internal Curing Using Perforated Cenospheres (2016), pp. 1–13

J. Wang, K. Zheng, N. Cui, X. Cheng, K. Ren, P. Hou, L. Feng, Z. Zhou, N. Xie, Green and durable lightweight aggregate concrete: the role of waste and recycled materials. Materials **13**(13), 3041 (2020). https://doi.org/10.3390/ma13133041

S. Weber, H.W. Reinhardt, A new generation of high performance concrete: concrete with autogenous curing. Adv. Cem. Based Mater. **6**, 59–68 (1997). https://doi.org/10.1016/s1065-7355(97)00009-6

Y. Wehbe, A. Ghahremaninezhad, Combined effect of shrinkage reducing admixtures (SRA) and superabsorbent polymers (SAP) on the autogenous shrinkage, hydration and properties of cementitious materials. Constr. Build. Mater. **138**, 151–162 (2017)

H.S. Wong, *Eco-Efficient Repair and Rehabilitation of Concrete Infrastructures*. Concrete with Superabsorbent Polymer (Elsevier Ltd., 2017) https://doi.org/10.1016/B978-0-08-102181-1.00017-4

P.P. Woyciechowski, M. Kalinowski, The influence of dosing method and material characteristics of Superabsorbent Polymers (SAP) on the effectiveness of the concrete internal curing. Materials **11**, 1600 (2018). https://doi.org/10.3390/ma11091600

N. Wu, M.A. Hubbe, O.J. Rojas, S. Park,. Permeation of polyelectrolytes and other solutes into the pore spaces of water-swollen cellulose: a review. Bioresources **4**(3), 1222–1262 (2009). https://doi.org/10.15376/biores.4.3.1222-1262

M. Wyrzykowski, S. Ghourchian, S. Sinthupinyo, N. Chitvoranund, T. Chintana, P. Lura, Internal curing of high performance mortars with bottom ash. Cem. Concr. Compos. **21**, 1–9 (2016). https://doi.org/10.1016/j.cemconcomp.2016.04.009

References

J. Xin, D. Huang, J. Li, C. Lin, Effect of SAP on the tensile creep of early-age concrete using ring specimens. Eur. J. Environ. Civ. Eng. **22**(7), 869–882 (2018). https://doi.org/10.1080/19648189.2016.1229223

N. Yadav, S.V. Deo, G.D. Ramtekka, Mechanism and benefits of internal curing of concrete using light weight aggregates and its future prospects in Indian construction. Int. J. Civil Eng. Technol. **8**(5), 323–334 (2017)

S. Yang, L. Wang, Effect of internal curing on characteristics of self-compacting concrete by using fine and coarse lightweight aggregates. J. Mater. Civ. Eng. **29**(10), 1–11 (2017). https://doi.org/10.1061/(ASCE)MT.1943-5533.0002044

L. Yang, X. Ma, J. Liu, X. Hu, Z. Wu, C. Shi, Improving the effectiveness of internal curing through engineering the pore structure of lightweight aggregates. Cem. Concr. Compos. **134**, 104773 (2022). https://doi.org/10.1016/j.cemconcomp.2022.104773

G. Ye, N.V. Tuan, H. Hao, in *Third International Conference on Sustainable Construction Materials and Technologies*. Rice Husk Ash (RHA) as Smart Materials to Mitigate Autogenous Shrinkage In Ultra High Performance Concrete (2013), p. 9

M.K. Yew, M.C. Yew, J.H. Beh, Effects of recycled crushed light expanded clay aggregate on high strength lightweight concrete. Mater. Int. **2**(3), 311–317 (2020)

V.Z. Zadeh, C.P. Bobko, Nano-mechanical properties of internally cured kenaf fiber reinforced concrete using nanoindentation. Cem. Concr. Compos. **52**, 9–17 (2014). https://doi.org/10.1016/j.cemconcomp.2014.04.002

M. Zaichenko, S. Lakhtaryna, A. Korsun, The influence of extra mixing water on the properties of structural lightweight aggregate concrete. Proc. Eng. **117**(1), 1036–1042 (2015). https://doi.org/10.1016/j.proeng.2015.08.228

J. Zhang, Y. Huang, K. Qi, Y. Gao, Interior relative humidity of normal-and high-strength concrete at early age. J. Mater. Civ. Eng. **24**(6), 615–622 (2012). https://doi.org/10.1061/(asce)mt.1943-5533.0000441

B. Zhang, C.S. Poon, Internal curing effect of high volume furnace bottom ash (FBA) incorporation on lightweight aggregate concrete. J. Sustain. Cem. Based Mater. **6**(6), 366–383 (2017). https://doi.org/10.1080/21650373.2017.1299053

Y.Z. Zhuang, D.D. Zheng, Z. Ng, T. Ji, X.F. Chen, Effect of lightweight aggregate type on early-age autogenous shrinkage of concrete. Constr. Build. Mater. **120**, 373–381 (2016). https://doi.org/10.1016/j.conbuildmat.2016.05.105

S. Zhutovsky, K. Kovler, A. Bentur, Efficiency of lightweight aggregates for internal curing of high strength concrete to eliminate autogenous shrinkage. Mater. Struct./materiaux Et Constr. **34**(246), 97–101 (2002). https://doi.org/10.1007/bf02482108

S. Zhutovsky, K. Kovler, A. Bentur, Assessment of water migration distance in internal curing of high-strength concrete. Am. Concr. Inst **220**, 181–200 (2004)

S. Zhutovsky, K. Kovler, A. Bentur, Influence of cement paste matrix properties on the autogenous curing of high-performance concrete. Cem. Concr. Compos. **26**(5), 499–507 (2004). https://doi.org/10.1016/S0958-9465(03)00082-9

S. Zhutovsky, K. Kovler, A. Bentur, Revisiting the protected paste volume concept for internal curing of high-strength concretes. Cem. Concr. Res. **41**(9), 981–986 (2011). https://doi.org/10.1016/j.cemconres.2011.05.007

S. Zhutovsky, K. Kovler, Influence of water to cement ratio on the efficiency of internal curing of high-performance concrete. Constr. Build. Mater. **144**, 311–316 (2017). https://doi.org/10.1016/j.conbuildmat.2017.03.203

S. Zhutovsky, K. Kovler, A. Bentur, Efficiency of lightweight aggregates for internal curing of high strength concrete to eliminate autogenous shrinkage. Mater. Struct. **35**(March), 97–98 (2002)

D. Zou, J. Weiss, Early age cracking behavior of internally cured mortar restrained by dual rings with different thickness. Constr. Build. Mater. **66**, 146–153 (2014). https://doi.org/10.1016/j.conbuildmat.2014.05.032

D. Zou, K. Li, W. Li, H. Li, T. Cao, Effects of pore structure and water absorption on internal curing efficiency of porous aggregates. Constr. Build. Mater. **163**, 949–959 (2018). https://doi.org/10.1016/j.conbuildmat.2017.12.170

Chapter 4
Physical, Deformation, Microstructure and Chemical Properties of Self Cured Agents and Concrete

4.1 Introduction

This chapter explores the physical properties, deformation characteristics, microstructure, interfacial transition zone (ITZ) and chemical properties of self cured concrete, offering further understanding of its performance and benefits. The primary physical properties discussed include density, water absorption, water desorption, shape and texture of the aggregates used for self curing agents. Lightweight aggregates (LWAs) exhibit a lower density compared to conventional aggregate due to the presence of water-retaining agents. Shrinkage is an essential parameter of self cured concrete. Self cured concrete exhibits lower autogenous and drying shrinkage compared to traditional concrete, as the internal curing process helps maintain moisture levels, reducing the potential for shrinkage-induced cracking. Additionally, self cured concrete shows reduced shrinkage and cracking due to the continuous supply of internal curing water, which mitigates the effects of drying concrete. The improved microstructure enhances the ITZ between the cement paste and aggregates, reducing porosity and increasing overall strength and durability.

4.2 Physical Properties of Lightweight Aggregate (LWA) as Self Curing Agents

Three (3) main properties of LWA as self curing agent namely water absorption, water desorption and shape and surface are further discussed in this chapter. The following sub-sections elaborate further the properties of LWA as self curing agents that affect the behavior of self curing concrete.

4.2.1 Water Absorption

Water absorption of porous aggregate is strongly related to its pore size distribution and porosity. The large pores in porous aggregate are rapidly saturated, especially larger than 1 μm pores, but have very poor capacity for water retention and will reduce the efficiency of internal curing (Ma et al. 2019; Zou et al. 2018). Besides, with reducing the pore sizes, water absorption becomes difficult. Even if a very small pore can absorb water in a vacuum, it is difficult to release water. It is assumed that 100 nm of pore size is a vital value for internal curing (Mechtcherine and Reinhardt 2012; Ma et al. 2019). Increasing the number of open pores is also very helpful for water absorption and internal curing in concrete. Usually, the high-strength of porous aggregate water absorption at a 1-h soak in water is less than 10%, but ordinary porous aggregate water absorption is around 15–20% (Wei et al. 2010). The explanation is possibly that cement particles is blocked pores of porous aggregate (Henkensiefken 2008). If air-dry porous aggregate is applied, water from the cement paste starts to be absorbed by the porous aggregates. Golias et al. (2012) discovered that oven-dry porous aggregate absorbed water from cement paste up to initial set, but its absorption of water is lower if the aggregates are soaked in water prior to mixing with cement. Castro et al. (2012) further discovered that only about 55% of the water from the aggregates if the latter is 24 h pre-wetted was absorbed. However, the amount of absorption would be increased up to 70–75% if the aggregate were firstly soaked for a few minutes in the water. Therefore, the pre-saturation is very necessary for porous aggregate. More water is absorbed when the volume of porous aggregate is higher and the internal curing effect is therefore, more noticeable. However, higher amount of porous aggregate will obviously reduce the strength of concrete due to its porous structure.

4.2.2 Water Desorption

The water desorption process is related to the pore size of porous aggregates and internal RH of concrete. Only water in pores larger than a certain size (critical pore) is released at constant RH (Nguyen and Le 2018; Henkensiefken 2008). Thus, the pores smaller than critical pore are still saturated. The effectiveness of internal curing depends on water desorption process of porous aggregate. Trtik et al. (2011) opined that three aspects should first be addressed in order to achieve a successful curing effect: (1) when the water is released by the self curing agent, (2) how much water can be released, (3) how far the internal curing water in hardened paste can go. The best curing efficiency can be achieved if internal curing water is released when adequate water can spread when it is needed. A majority of water in the pores of porous aggregate is released at high RH (> 96%) (Henkensiefken 2008). When RH is reduced to 92%, almost all the water in the pores is released.

4.2.3 Shape and Surface Texture

The shape and texture of an aggregate have a more significant effect on the qualities of freshly mixed concrete than on hardened concrete. Elongated, angular and rough surface textured aggregates require significantly more water to form workable concrete than rounded, smooth and compact aggregates (Jagadish and Jagadeesan 2015; Kosmatka and Wilson 2011). As a result, angular aggregate particles require additional cement to retain the same water ratio to cementing materials. However, with up to standard gradation, both crushed and uncrushed aggregates (of the same aggregate type) often provide the same strength for the same cement factor. The amount of water and cement required for mixing tends to increase as the void content of the aggregate increases (Tang et al. 2017). Besides, voids between aggregate particles increase with aggregate angularity (Kosmatka and Wilson 2011). The binding between cement pastes and a particular aggregate often increases as the particles get rougher and more angular. This increase in bond is considered when aggregates for concrete that require strong flexural strength or compressive strength. Yehia et al. (2014) reported that flexural strength improved by around 80%, attributed to the enhanced bond between the aggregate and cement paste caused by the angular particle shape and crushed surface.

Aggregate particles that are flat or elongated should be avoided or limited to no more than 15% of the total aggregate mass. According to ASTM D4791-19 (2023), particles are considered flat and elongated when the ratio of length to thickness exceeds a pre-determined threshold. Due to the increased mixing water required by such aggregate particles, the strength of concrete, particularly in flexure, may be compromised if the water-cement ratio is not adjusted (Kosmatka et al. 2002).

4.3 Autogeneous Shrinkage and Drying Shrinkage of Self Cured Concrete

Cracking is a common occurrence in cement-based materials. Cracking has a detrimental effect on the structural integrity of the concrete. Generally, it is believed to be related to shrinkage. Chemical shrinkage and autogenous shrinkage are responsible for the cracking of high-performance cement-based materials (Craeye et al. 2011; Cusson and Hoogeveen 2008). Additionally, the capillary porosity created by chemical shrinkage is requisite for autogenous shrinkage (Lura et al. 2003). During the hardening process in concrete, both autogenous shrinkage and chemical shrinkage occur due to cement hydration. The absolute volume change corresponds to the chemical shrinkage, whereas the apparent volume change corresponds to the autogenous shrinkage during cement hydration (Jensen and Hansen 2001). Deformation of cement-based materials caused by the loss of capillary water is referred to as drying shrinkage. In order to mitigate autogenous shrinkage caused by chemical shrinkage pores, additional water must be introduced to fill the capillary porosity created by the chemical shrinkage (Ma et al. 2015).

Pre-wetted porous aggregate can more effectively minimize autogenous shrinkage compared to the oven-dry porous aggregate (Ghourchian et al. 2013; Bentur et al. 2001; Lura et al. 2004; Liu et al. 2019; Ji et al. 2015). This is because of the dry porous aggregate's inability to absorb sufficient additional water from concrete during the plastic stage (Castro et al. 2012). Dry porous aggregate can achieve almost the same result as pre-wetted porous aggregate if it can absorb the required amount of extra water from the concrete mixture (Golias 2012). Pre-wetted porous aggregate gradually released internal curing water, extending the time that internal RH remains at 100%. As a result, this reduces autogenous shrinkage whereby refining pore structure and improving the hydration (Ma et al. 2019). Besides, the kinds of LWA might have an impact on their efficacy. Zhuang et al. (2016) discovered that pre-wetted porous aggregate with higher water absorption and lower strength reduces autogenous shrinkage. However, porous aggregate with higher strength and lower water absorption will have a better effect if put in the dry condition. In general, porous aggregate minimizes the total shrinkage at 28 days (Henkensiefken 2008) while nevertheless increasing drying shrinkage (Costa et al. 2012). According to Costa et al. (2012), a 34% increase in drying shrinkage was observed in the specimen consisting of fine pumice aggregate compared to the reference specimen with conventional aggregates. The increase in drying shrinkage is related to lower elastic modulus and increased water binder ratio (Lura and Bisschop 2004; Zhutovsky and Kovler 2012).

The development rate in autogenous shrinkage of HPC can be minimized by adding SAP. The increase in the content of SAP, regardless of size, reduced the autogeneous shrinkage significantly (Liu et al. 2021; Mechtcherine et al. 2021). The HPC internally cured with SAPs showed a lower autogeneous shrinkage than that without SAPs, which was due to the decreased self-dessication of HPC internally cured with SAPs (Shen 2020). Thus, the emergence of cracking can be delayed (Huang and Wang 2012). The dosage (Jensen and Hansen 2002; Huang and Wang 2012; Esteves 2009), type (Oh and Choi 2018; Sarbapalli et al. 2016; Jensen and Hansen 2001, 2002), particle size (Jensen and Hansen 2002; Huang and Wang 2012; Esteves 2009), and water saturated state of SAPs (Igarashi and Watanabe 2006) plays an important role in the efficiency of self curing. Shen et al. (2016) reported that the autogenous shrinkage of self cured concrete decreased with increase of internal curing water provided by SAP. The rate of autogenous shrinkage reduced as the concrete ages increased, while the autogenous shrinkage rates of all mixtures were remarkably similar at 28 days. The effect of self curing with varied dosages of SAP on autogenous shrinkage can be related to the extra water supplied by SAP (Craeye et al. 2011). The water inside the SAP acts as the water contained inside pores at the time of batching, and it is available to assist internal curing while not affecting the initial w/c (Browning et al. 2011; Wang et al. 2009). Internal RH decreases as a result of water loss by self-desiccation during the hydration process, resulting in driving force (Jensen and Hansen 2001). Thus, internal curing water is released from SAP into the cement paste, contributing to continuing hydration of cement. A higher

4.3 Autogeneous Shrinkage and Drying Shrinkage of Self Cured Concrete

dosage of SAP depicts more water being released when autogenous shrinkage is reduced. The drying shrinkage also reduced with the increase of dosage of SAP as reported by Jensen and Hansen (2002), Ma et al. (2019), Assmann and Reinhardt (2014) and Kong et al. (2014). However, if the environment is arid, the water from the SAP particles evaporates quickly, resulting in cracking. The additional water absorbed by the self curing agent has a considerable impact on the drying shrinkage of self cured cement-based material.

Using a chemical curing agent such as PEG was also reported to reduce early age shrinkage cracks (Sathanandham et al. 2013; Amin et al. 2021; Bashandy et al. 2017). Amin et al. (2021) investigated engineering properties on self curing concrete by using polyethylene glycol. The results revealed that adding PEG to the concrete mixes to perform the self curing role contributes to reducing dry shrinkage compared with reference concrete. Hence, the application of the self curing explained in this research increased the subsequent degree of hydration of cement and the chemical shrinkage, consequently effectively reducing early and late shrinkage. However, only a few researchers had investigated the autogenous and drying shrinkage of concrete when PEG was used as a curing agent.

Bentur et al. (2001) discovered that when the amount of fine LWA replacing normal-weight aggregate reached 25%, autogenous shrinkage of HPC could be eliminated. However, Geiker et al. (2004) reported that 20% replacement level of normal weight aggregate with LWA could eliminate autogenous shrinkage of HPC. The optimum amount of LWA to replace normal aggregate is determined by the pore structure and water absorption rate of the LWA. In addition, vacuum-saturated LWA significantly reduces autogenous shrinkage compared to those of oven-dry LWA. The effect of using air dry and SSD condition of LWA toward autogeneous shrinkage is depicted in Fig. 4.1.

This is because dry LWA is incapable of absorbing sufficient additional water from concrete during the plastic stage (Castro et al. 2012). If dry LWA absorbs the required amount of internal curing water from the concrete mixture, it is similar to pre-wetted LWA (Golias et al. 2012). By optimizing the LWA porosity and particle size, it is possible to achieve an effective internal curing condition while using the least LWA (Kovler et al. 2004; Zhutovsky et al. 2002). Zhutovsky et al. (2004) reported that using fine LWA was more effective than coarse LWA at ensuring adequate distribution of internal curing water in concrete. When the particle size of LWA was reduced, it distributed more uniformly in the matrix and improved the internal curing of the cement paste. Besides, the amount of internal water absorbed by LWA and the desorption process of LWA effect on shrinkage of high-performance cement-based materials. In general, pre-wetting LWA for 24 h significantly reduces and even eliminates autogenous shrinkage (Akcay and Tasdemir 2010). Furthermore, the type of LWA used can have an effect on its effectiveness. Suppose the water to cement ratio is maintained constant. In that case, the LWA with lower strength and a high-water absorption rate have a better inhibitory effect on autogenous shrinkage in pre-wetted conditions. In contrast, LWA with higher strength and low-water absorption rate results better when added in the dry state (Zhuang et al. 2016).

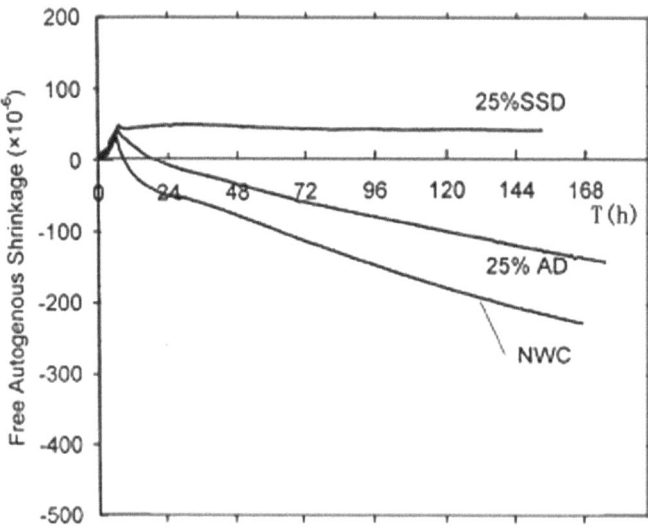

Fig. 4.1 Effect of autogenous shrinkage of concrete with air-dry (AD) and saturated-surface-dry (SSD) lightweight coarse aggregate (Bentur et al. 2001, Copyright 2001, with permission from Elsevier)

According to Dong et al. (2009), even as increasing the moisture content of expanded shale may slow shrinkage, drying shrinkage cannot be completely avoided by increasing the amount of LWA. However, Deboodt et al. (2016) believed that fine LWA had little benefit in reducing drying shrinkage, especially when an external wet curing period of three days or longer was used. When the environment is dry, a noticeable moisture gradient in the concrete and the drying of the concrete surface may occur (Han et al. 2013). The underlying reason for shrinkage during the plastic stage is the rapid loss of water from the concrete surface due to evaporation and the physical process of capillary pressure build-up (Boshoff and Combrinck 2013; Cohen et al. 1990). While the mechanism of dry shrinkage is similar to that of plastic shrinkage, the former is believed to occur more frequently during the hardening stage. Autogenous shrinkage of high-performance cement-based materials occurs due to the effect of a significant decrease in RH within the pore system and an increase in capillary stress (Ghafari et al. 2016). Depending on the water content of the system and the pore size distribution caused by hydration, chemical shrinkage is transformed into autogenous shrinkage (Mechtcherine and Reinhardt 2012).

Abadel (2023) discovered that increasing the percentage of LWAs reduced the shrinkage of the ultra-high performance concrete (UHPC). Several factors contributed to the decrease in autogenous shrinkage of ultra-high performance concrete (UHPC) as the proportion of LWAs increased. Additionally, the incorporation of LWAs into the UHPC mixture improves internal drainage, allowing excess water to flee more rapidly and thereby decreasing the concrete's overall shrinkage. The enhanced drainage is attributable to the larger pore size of the LWA, which

4.4 Microstructure Examination Using Scanning Electron Microscope (SEM)

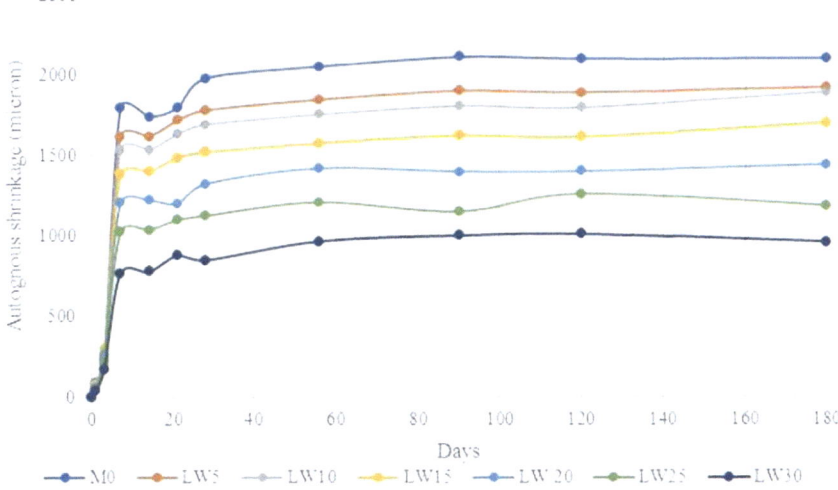

Fig. 4.2 Effect of LWA on shrinkage of UHPC specimens (Abadel 2023)

facilitates more efficient conveyance of excess water compared to the smaller pore size of the cement paste. By means of improved drainage, the quantity of internal moisture accessible for self-desiccation can be further diminished, thereby mitigating the potential for fractures (Akeed et al. 2022). The findings of the study indicated that the sample containing 0% LWA experienced the highest shrinkage (2100 μm) throughout all curing days. In contrast, the sample containing 30% LWA exhibited the least shrinkage (958 μm) throughout all curing days as shown in Fig. 4.2. This demonstrates the advantageous impact of LWAs on the autogenous shrinkage of UHPC and offers significant insights into the potential use of LWAs as a partial replacement for fine aggregates in UHPC in order to enhance its characteristics.

4.4 Microstructure Examination Using Scanning Electron Microscope (SEM)

The effect of aggregate particles on the concrete and mortar microstructure continues to be of significant interest to researchers, as it influences the mechanical strength and durability of the structure (Wong and Buenfeld 2006). The strength of the bond between the aggregates and the cement paste matrix, a zone called the interfacial transition zone (ITZ), which affected by the morphology of the aggregates, including the size, texture, shape, roughness, and porosity of the aggregates (Hilal 2016). Between cement paste and conventional aggregate, ITZ has a higher porosity and contains more calcium hydroxide (Ca $(OH)_2$) and ettringite (Mehta and Monteiro 2001). However, a thin and denser ITZ was formed between porous aggregate and cement paste matrix (Elsharief et al. 2005; Liu et al. 2011, 2019). In addition, the formation of calcium silicate hydrate (C–S–H) in high quantities (Agostini et al. 2010; Al Saffar et al. 2019; Suwan and Wattanachai 2017; Suzuki et al. 2009) and the homogeneousness (Zhang and Gjerv 1990; Larbi 1993) are also improved in ITZ. Thus, enhance the strength of internally curing concrete and establishes a well-bonding between porous aggregate and cement paste at ITZ. Several techniques were employed to verify the mineralogical change in self curing concrete that shows the formation of calcium silicate hydrate as discussed in the following paragraph.

Along with the physical and mechanical qualities, a study of the microstructure using SEM is required to determine the type, amount, size, shape and distribution of phases present in a solid concrete formed under different curing conditions (Madduru et al. 2018). The ITZ of concrete samples was examined to determine the distribution of atomic ratios in aggregates and hydrated phases of ITZ. Limited research was done on the effect of microstructure examination on the concrete containing LWA as a self curing agent. The investigation was carried out by Suwan and Wattanachai (2017) which makes the comparison between the results of improvement of strength in hardened concrete and aggregate-cement bonding. It was found that a well bonding between autoclaved aerated lightweight concrete (AAC-LWA) and cement paste at their ITZ whereby AAC-LWA replacement levels adopted were from 20 to 40% as shown in Fig. 4.3. By this, it can be explained that by the suitable amount AAC-LWA, it served the additional water for internal curing whereby extra water in aggregates transferred to the cement paste across ITZ. As a result, increasing the hydration level that contributes to the enhancement of the cement binders and the uniformity in ITZ. Additionally, these factors were the key reasons that strengthened the concrete. Sun et al. (2015) applied SEM technique to compute the ITZ microstructure in cement mortar containing LWA with 0.35 water to cement ratio with 50% of LWA volume fraction. The results indicated that the internal curing reduces the permeability of the ITZ section due to less percolated porosity and smaller characteristic pore sizes compared with the samples without internal curing.

4.5 Effect of Curing Agent on Interfacial Transition Zone (ITZ)

The interfacial transition zone (ITZ) between cement pastes and aggregate is the most important interface in concrete. There is high porosity and more calcium hydroxide (Ca $(OH)_2$) and ettringite in ITZ between cement paste and conventional aggregate (Joudah et al. 2021). However, a thin and denser ITZ was formed between porous aggregate and cement paste matrix (Liu et al. 2011, 2019; Elsharief et al. 2005; Sun et al. 2015) as depicted in Fig. 4.4.

The formation of C–S–H in high quantities (Al Saffar et al. 2019; Suzuki et al. 2009; Suwan and Wattanachai 2017; Agostini et al. 2010) and the homogeneousness (Al-Fasih et al. 2021; Zhang and Gjerv 1990) are also improved in ITZ.

4.5 Effect of Curing Agent on Interfacial Transition Zone (ITZ)

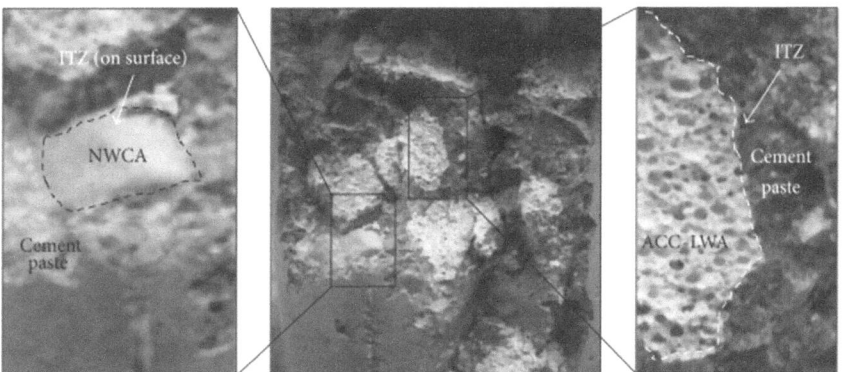

Fig. 4.3 Normal-bonded (left) and well-bonded AAC-LWA (right) with cement paste on interfacial transition zone (ITZ) (Suwan and Wattanachai 2017)

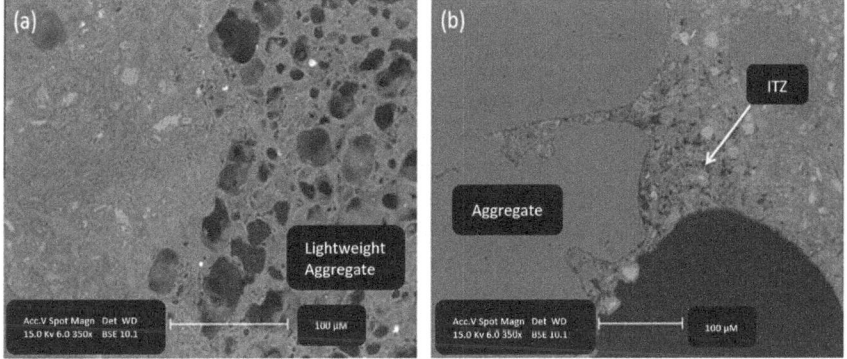

Fig. 4.4 ITZ microstructure in cement mortar **a** with porous aggregate, **b** with conventional aggregate (Reprinted from Sun et al. 2015, Copyright 2015, with permission from Elsevier)

Thus, it enhances the strength of internally curing concrete and establishes a well bonding (Suwan and Wattanachai 2017) between porous aggregate and cement paste at ITZ. Moreover, some hydrated products that were discovered in pores of porous aggregate contributed to an increase in strength at later ages (Al Saffar et al. 2019; Holm et al. 2003; Bentz and Stutzman 2006). Sun et al. (2015) applied SEM techniques to compute the ITZ microstructure, founded a 3D digital model of ITZ using 3D image reconstruction techniques, and applied the mesoscale chemo thermal hydraulic model to simulate the development of ITZ. The research revealed that internal curing can improve the durability of concrete. It is proposed that multi-physical hydration models be linked to microstructure characterization and transport properties in order to examine ITZ microstructure development.

The additional of SAP could also enhance the degree of hydration and densify the cement matrix on the surrounding SAP (Wang et al. 2016; Yang et al. 2019). However,

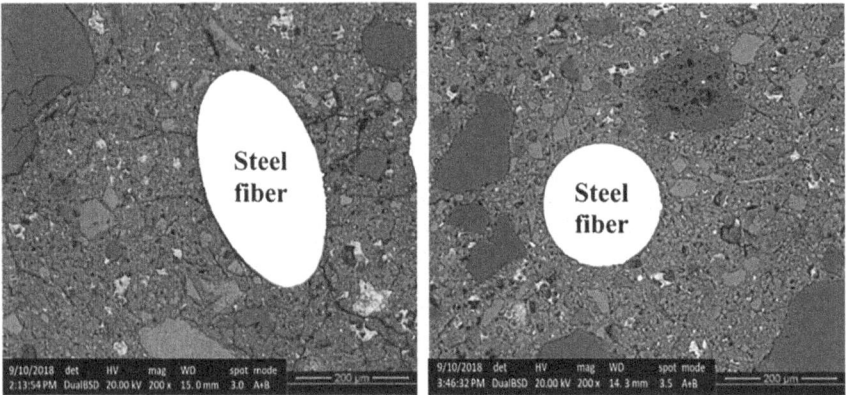

Fig. 4.5 The microstructure of the high-performance concrete with SAP (Reprinted from Liu et al. 2020, Copyright 2020, with permission from Elsevier)

the water release from SAP can leave pores (Mechtcherine et al. 2021; Ma et al. 2015) and undermine the properties of the cement matrix (Kong et al. 2014). Studies revealed by Lui et al. (2020) reported that the development of pores can impair the ITZ and bond strength between SAP and the cement matrix as shown in Fig. 4.5. This effect increased as the SAP content and particle size of the SAP increased. Ridi et al. (2011) found that the incorporation of SAP into high-performance concrete (HPC) helps the densification of the microstructure of the concrete. They also discovered via experimentation that the self curing procedure is associated with increased strength, a higher degree of hydration, and penetration resistance in heat-cured concrete, as well as a better microstructure with a lower porosity in HPC.

Microscopic investigation of normal strength concrete containing LWA acts as a curing agent under external and without external curing is an important parameters to assess the effectiveness of self curing actions in concrete. Tuan et al. (2011), Suwan and Wattanachai (2017) and Kim and Lee (2018) have proved that the internal microstructure of high-performance concrete containing LWA as a curing agent was enhanced attributed to improving the interfacial transition zone (ITZ) due to more development of calcium-silicate hydrate (C–S–H) gel in self cured concrete compared to those control concrete specimens.

4.6 Chemical Analysis Using X-Ray Diffraction (XRD)

Madduru et al. (2018) investigated chemical curing agent compounds to improve the curing efficiency in concrete. It was revealed that the use of hydrophilic and hydrophobic chemicals resulted in a dense microstructure with a Ca/Si = 1.12 and 1.37, respectively. This is an indication of stable C–S–H gel formation compared to no cured specimens as shown in Fig. 4.6.

4.6 Chemical Analysis Using X-Ray Diffraction (XRD)

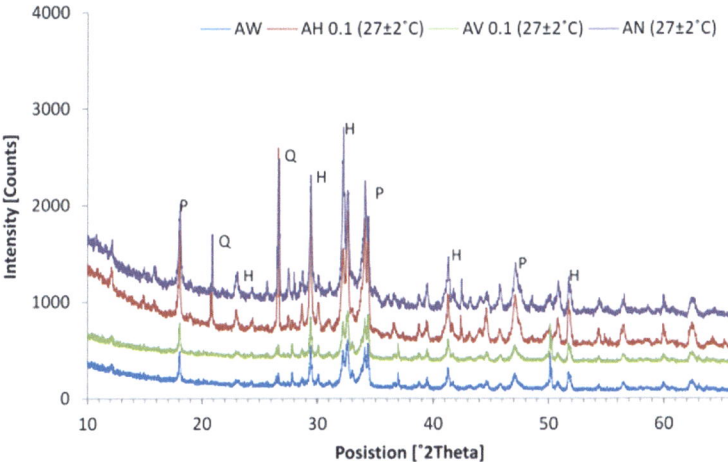

Fig. 4.6 X-ray diffraction peaks of SCC subjected to different curing regimes at 27 ± 2 °C (used with permission of Madduru et al. 2018; permission conveyed through Copyright Clearance Center, Inc.)

While no cured specimens exhibited poor microstructure with interlinking of micro cracks and ettringite formation. Tang (2017) reported that the C–H content of concrete containing oven dried LWA was the lowest. This is due to the water absorption characteristics of the oven dried LWAs, resulting in free water reduction and lower rates of hydration. As for concrete containing with LWA pre-soaked in water for 1 h, its C–H content is higher than that of concrete containing oven dried LWA as shown in Fig. 4.7.

Abadel (2023) discovered the first XRD spectrum as depicted in Fig. 4.8, the UHPC sample with 0% pumice replacement reveals high peaks of portlandite ($Ca(OH)_2$), calcium-silicate-hydrate (C–S–H) and ettringite strengths. It appears that the UHPC sample without fine aggregate replacement has higher crystallinity. Sharp and powerful spectrum peaks may suggest highly crystalline phases like portlandite, a common cement-based phase (Lim et al. 2019). The UHPC sample with 15% pumice replacement of fine aggregates showed reduced peak portlandite, C–S–H, and ettringite intensities in the second XRD spectrum. Pumice replacing fine aggregates may have reduced UHPC crystallinity. Some of the same peaks as the first spectrum suggest that the 15% replacement UHPC sample has some of the same crystalline phases as the 0% replacement sample. Portlandite, calcium-silicate-hydrate, and ettringite intensities are considerably lower in the third XRD spectrum, which represents the UHPC sample with 30% pumice replacement of fine particles. Replacing fine particles with pumice may have further reduced UHPC crystallinity (Akeed et al. 2022).

In summary, the presence of internal curing agents ensures a more complete hydration of the cement particles, leading to a more stable and homogeneous microstructure and chemical composition. This results in higher cement paste quality and

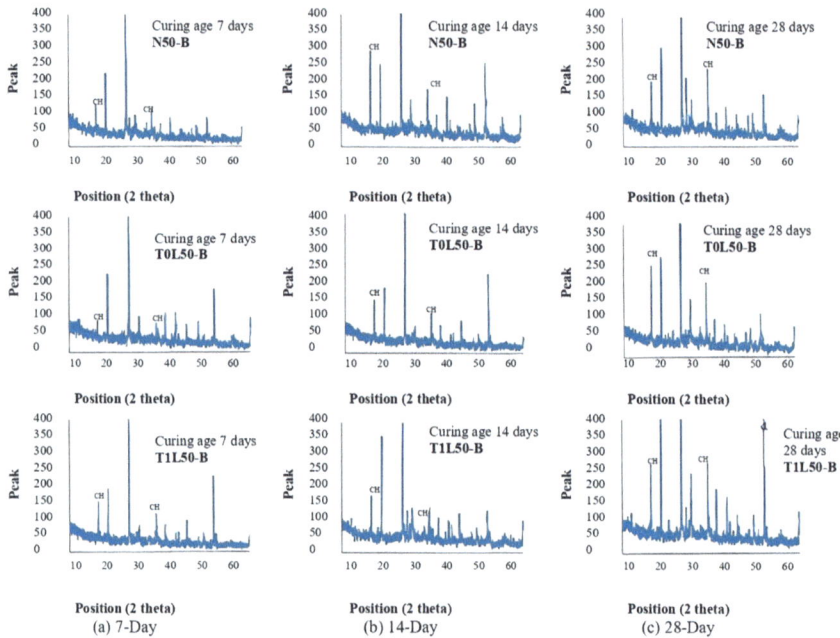

Fig. 4.7 X-ray diffraction analysis (Tang 2017)

increased formation of hydration products like calcium silicate hydrate (C–S–H), which contribute to the concrete's strength and durability. Self cured concrete offers several advantages over conventional concrete, including enhanced physical properties, improved deformation characteristics, refined microstructure and superior chemical properties. The incorporation of internal curing agents ensures a more complete and uniform hydration process, leading to better performance and longevity. Understanding these aspects is essential for optimizing the use of self cured concrete in various construction applications, ultimately contributing to more durable and sustainable infrastructure.

4.6 Chemical Analysis Using X-Ray Diffraction (XRD)

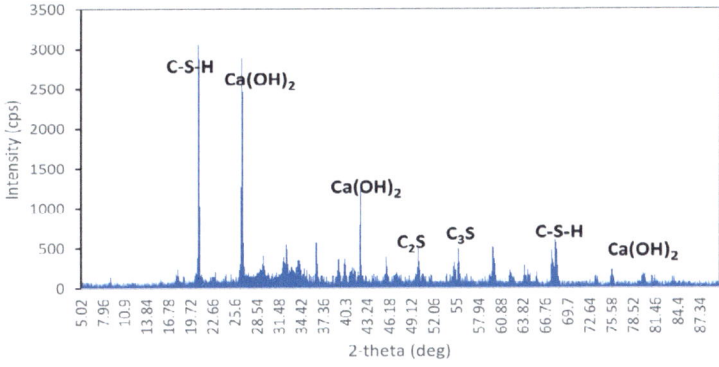

a) First XRD spectrum (LW0)

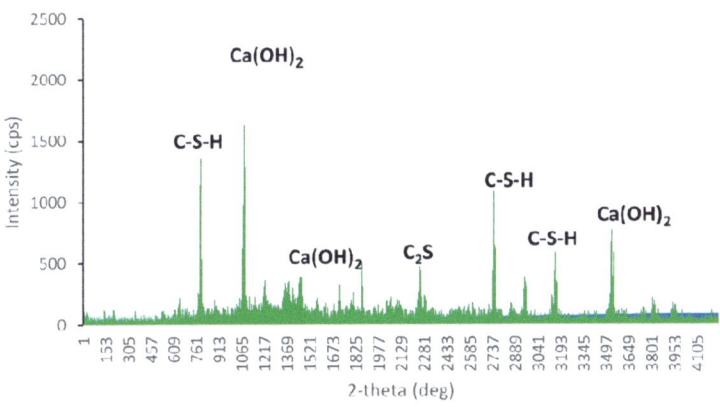

b) Second XRD spectrum (LW15)

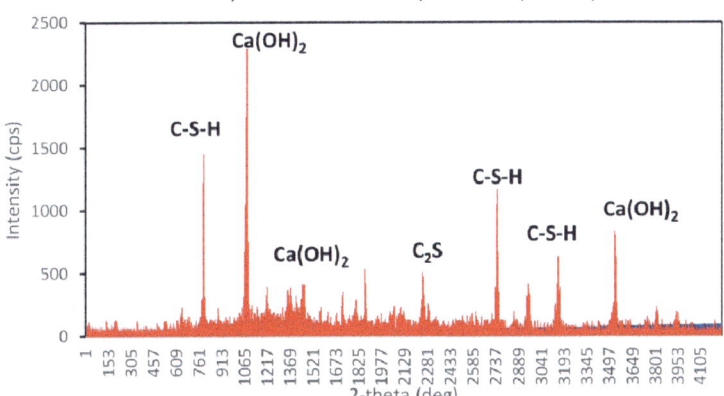

c) Third XRD spectrum (LW30)

Fig. 4.8 XRD spectra of UHPC: **a** LW0, **b** LW15, **c** LW30 (Abadel 2023)

References

A.A. Abadel, Physical, mechanical and microstructure characteristics of ultra-high-performance concrete containing lightweight aggregates. Materials **16**(13) (2023). https://doi.org/10.3390/ma16134883

F. Agostini, C. Davy, F. Skoczylas, T. Dubois, Effect of microstructure and curing conditions upon the performance of a mortar added with Treated Sediment Aggregates (TSA). Cem. Concre. Res. **40**, 1609–1619 (2010). https://doi.org/10.1016/j.cemcon-res.2010.07.003

B. Akcay, M.A. Tasdemir, Effects of distribution of lightweight aggregates on internal curing of concrete. Cem. Concr. Compos. **32**(8), 611–616 (2010). https://doi.org/10.1016/j.cemconcomp.2010.07.003

M.H. Akeed, S. Qaidi, H.U. Ahmed, R.H. Faraj, A.S. Mohammed, W. Emad, B.A. Tayeh, A.R. Azevedo, Ultra-high-performance fiber-reinforced concrete. Part II: hydration and microstructure. Case Stud. Constr. Mater. **17**(June), e01289 (2022). https://doi.org/10.1016/j.cscm.2022.e01289

M.Y. Al-Fasih, G.F. Huseien, I.S.B. Ibrahim, A.R. Sam, H.A. Algaifi, R. Alyousef, Synthesis of rubberized alkali-activated concrete: experimental and numerical evaluation. Constr. Build. Mater. **303**, 124526 (2021)

D.M. AlSaffar, A,J, Al Saad, B.A. Tayeh, Effect of internal curing on behavior of high performance concrete: an overview. Case Stud. Constr. Mater. **10**, e00229 (2019). https://doi.org/10.1016/j.cscm.2019.e00229

M. Amin, A.M. Zeyad, B.A. Tayeh, I.S. Agwa, Engineering properties of self-cured normal and high strength concrete produced using polyethylene glycol and porous ceramic waste as coarse aggregate. Constr. Build. Mater. **299**, 124243 (2021). https://doi.org/10.1016/j.conbuildmat.2021.124243

A. Assmann, H. Reinhardt, Tensile creep and shrinkage of SAP modified concrete. Cem. Concr. Res. **58**, 179–185 (2014). https://doi.org/10.1016/j.cemconres.2014.01.014

ASTM D4791-19, *Standard Test Method for Flat Particles, Elongated Particles, or Flat and Elongated Particles in Coarse Aggregate* (2023)

A.A. Bashandy, N.N. Meleka, M. Hamad, Comparative study on the using of PEG and PAM as curing agents for self-curing concrete. Chall. J. Concr. Res. Lett. **8**, 1 (2017). https://doi.org/10.20528/cjcrl.2017.01.001

A. Bentur, S.I. Igarashi, K. Kovler, Prevention of autogenous shrinkage in high-strength concrete by internal curing using wet lightweight aggregates. Cem. Concr. Res. **31**, 1587–1591 (2001). https://doi.org/10.1016/s0008-8846(01)00608-1

D.P. Bentz, P.E. Stutzman, Curing, hydration and microstructure of cement paste. ACI Mater. J. **103**, 348–356 (2006). https://doi.org/10.14359/18157

W.P. Boshoff, R. Combrinck, "Modelling the severity of plastic shrinkage cracking in concrete. Cem. Concr. Res. **48**, 34–39 (2013). https://doi.org/10.1016/j.cemconres.2013.02.003

J. Browning, D. Darwin, D. Reynolds, B. Pendergrass, Lightweight aggregate as internal curing agent to limit concrete shrink- age. ACI Mater. J. **108**, 638–644 (2011)

J. Castro, I. de La Varga, J. Weiss, Using isothermal calorimetry to assess the water absorbed by fine LWA during mixing. J. Mater. Civil Eng. **24**, 996–1005 (2012). https://doi.org/10.1061/(asce)mt.1943-5533.0000496

H. Costa, E. Júlio, J. Lourenço, New approach for shrinkage prediction of high-strength lightweight aggregate concrete. Constr. Build. Mater. **35**, 84–91 (2012). https://doi.org/10.1016/j.conbuildmat.2012.02.052

M.D. Cohen, J. Olek, W.L. Dolch, Mechanism of plastic shrinkage cracking in portland cement and portland cement-silica fume paste and mortar. Cem. Concr. Res. **20**(1), 103–119 (1990)

B. Craeye, M. Geirnaert, G. de Schutter, Super absorbing polymers as an internal curing agent for mitigation of early-age cracking of high-performance concrete bridge decks. Constr. Build. Mater. **25**, 1–13 (2011). https://doi.org/10.1016/j.conbuildmat.2010.06.063

References

D. Cusson, T. Hoogeveen, Internal curing of high-performance concrete with pre-soaked fine lightweight aggregate for prevention of autogenous shrinkage cracking. Cem. Concr. Res. **38**(6), 757–765 (2008). https://doi.org/10.1016/j.cemconres.2008.02.001

T. Deboodt, F. Tengfei, H.I. Jason, Evaluation of FLWA and SRAs on autogenous deformation and long-term drying shrinkage of high performance concrete. Constr. Build. Mater. **119**, 53–60 (2016). https://doi.org/10.1016/j.conbuildmat.2016.05.068

S. Dong, B. Zhang, Y. Ge, J. Yuan, Influence of lightweight aggregate on shrinkage reducing efficiency of concrete. Kuei Suan Jen Hsueh Pao/j. Chin. Ceram. Soc. **37**(3), 465–469 (2009)

A. Elsharief, M.D. Cohen, J. Olek, Influence of lightweight aggregate on the microstructure and durability of mortar. Cem. Concr. Res. **35**, 1368–1376 (2005). https://doi.org/10.1016/j.cemconres.2004.07.011

L.P. Esteves, *Internal Curing in Cement-Based Materials*. Ph.D. Thesis, Aveiro, Portugal (2009)

E. Ghafari, A.G. Seyed, H. Costa, J. Eduardo, P. Antonio, D. Luisa, Effect of supplementary cementitious materials on autogenous shrinkage of ultra-high performance concrete. Constr. Build. Mater. **127**, 43–48 (2016). https://doi.org/10.1016/j.conbuildmat.2016.09.123

M.R. Geiker, D.P. Bentz, O.M. Jensen O. M, *Mitigating Autogenous Shrinkage by Internal Curing* (American Concrete Institute, ACI Special Publication, SP-218, 2004), pp. 143–154

S. Ghourchian, M. Wyrzykowski, P. Lura, M. Shekarchizadeh, B. Ahmadi, An investigation on the use of zeolite aggregates for internal curing of concrete. Constr. Build. Mater. **40**, 135–144 (2013). https://doi.org/10.1016/j.conbuildmat.2012.10.009

M. Golias, J. Castro, W. Weiss, The influence of the initial moisture content of lightweight aggregate on internal curing. Constr. Build. Mater. **35**, 52–62 (2012). https://doi.org/10.1016/j.conbuildmat.2012.02.074

Y. Han, J. Zhang, Z. Wang, Influence of pre-wetted lightweight aggregate on early-age shrinkage of high strength concrete. Kuei Suan Jen Hsueh Pao/j. Chin. Ceram. Soc. **41**(8), 1070–1078 (2013). https://doi.org/10.7521/j.issn.0454-5648.2013.08.08

R. Henkensiefken, *Internal Curing in Cementitious Systems Made with Saturated Lightweight Aggregate*. Ph.D. Thesis (Purdue University, West Lafayette, IN, USA, 2008)

A.A. Hilal, in *High Performance Concrete Technology and Applications*. Microstructure of Concrete (2016), (pp. 3–24). https://doi.org/10.5772/64574

T.A. Holm, O.S. Ooi, T.W. Bremner, in *Proceedings of the Theodore Bremner Symposium on High-Performance Lightweight Concrete, Sixth CANMET/ACI International Conference on Durability of Concrete, Thessalonikē, Greece, 1–7 June 2003*. Moisture Dynamics in Lightweight Aggregate and Concrete (2003), pp. 167–184

Z. Huang, J. Wang, Effects of SAP on the performance of UHPC. Bull. China Ceram. Soc. **21**, 539–544 (2012)

S. Igarashi, A. Watanabe, in *Proceedings of the International RILEM Conference*. Experimental Study on Prevention of Autogenous Deformation by Internal Curing Using Super-Absorbent Polymer Particles (2006), pp. 77–86

V. Jagadish, R. Jagadeesan, A feasibility study on artificial aggregates using waste materials. J. Civil Eng. Environ. Technol. **2**(3), 292–296 (2015)

O.M. Jensen, P.F. Hansen, Water-entrained cement-based materials: I. Principles and theoretical background. Cem. Concr. Res. **31**, 647–654 (2001)

O.M. Jensen, P.F. Hansen, Water-entrained cement-based materials: II. Experimental observations. Cem. Concr. Res. **32**, 973–978 (2002). https://doi.org/10.1016/s0008-8846(02)00737-8

T. Ji, D.D. Zheng, X.F. Chen, X.J. Lin, H.C. Wu, Effect of prewetting degree of ceramsite on the early-age autogenous shrinkage of lightweight aggregate concrete. Constr. Build. Mater. **98**, 102–111 (2015). https://doi.org/10.1016/j.conbuildmat.2015.08.102

Z.H. Joudah, G.F. Huseien, M. Samadi, N.H.A.S. Lim, Sustainability evaluation of alkali-activated mortars incorporating industrial wastes. Mater. Today Proc. **46**, 1971–1977 (2021). https://doi.org/10.1016/j.matpr.2021.02.454

H.K. Kim, H. Lee, Hydration kinetics of high-strength concrete with untreated coal bottom ash for internal curing. Cem. Concr. Compos. **91**, 67–75 (2018). https://doi.org/10.1016/j.cemconcomp.2018.04.017

X.M. Kong, Z.L. Zhang, Z.C. Lu, Effect of pre-soaked superabsorbent polymer on shrinkage of high-strength concrete. Mater. Struct. **48**, 2741–2758 (2014). https://doi.org/10.1617/s11527-014-0351-2

S.H. Kosmatka, B. Kerkhoff, W.C. Panarese, *Design and Control of Concrete Mixtures* (Fourteen E. Portland Cement Association, 2002)

S. Kosmatka, M.L. Wilson, Design and control of concrete mixtures: The guide to applications, methods, and materials. Portl. Cem. Assoc. (2011). https://doi.org/10.1007/springerreference_5435

K. Kovler, A. Souslikov, A. Bentur, Pre-soaked lightweight aggregates as additives for internal curing of high-strength concretes. Cem. Concr. Aggreg. **26**(2), 131–138 (2004). https://doi.org/10.1520/cca12295

J. Larbi, Microstructure of the interfacial zone around aggregate particles in concrete. HERON (1993)

J.L.G. Lim, S.N. Raman, M. Safiuddin, M.F.M. Zain, R. Hamid, Autogenous shrinkage, microstructure and strength of ultra-high performance concrete incorporating carbon nanofiber. Materials **12**(320) (2019). https://doi.org/10.3390/ma12020320

X. Liu, K.S. Chia, M. Zhang, Water absorption, permeability and resistance to chloride-ion penetration of lightweight aggregate concrete. Constr. Build. Mater. **25**(1), 335–343 (2011). https://doi.org/10.1016/j.conbuildmat.2010.06.020

J. Liu, C. Shi, N. Farzadnia, X. Ma, Effects of pretreated fine lightweight aggregate on shrinkage and pore structure of ultra-high strength concrete. Constr. Build. Mater. **204**, 276–287 (2019). https://doi.org/10.1016/j.conbuildmat.2019.01.205

J. Liu, N. Farzadnia, C. Shi, Effects of superabsorbent polymer on interfacial transition zone and mechanical properties of ultra-high performance concrete. Constr. Build. Mater. **231**, 117142 (2020). https://doi.org/10.1016/j.conbuildmat.2019.117142

J. Liu, N. Farzadnia, K.H. Khayat, C. Shi, Effects of SAP characteristics on internal curing of UHPC matrix. Constr. Build. Mater. **280**, 122530 (2021). https://doi.org/10.1016/j.conbuildmat.2021.122530

P. Lura, O. Mejlhede, K.V. Breugel, Autogenous shrinkage in high-performance cement paste: an evaluation of basic mechanisms. Cem. Ad Concr. Res. **33**, 223–232 (2003)

P. Lura, J. Bisschop, On the origin of eigenstresses in lightweight aggregate concrete. Cem. Concr. Compos. **26**, 445–452 (2004). https://doi.org/10.1016/s0958-9465(03)00072-6

P. Lura, D.P. Bentz, D.A. Lange, K. Kovler, A. Bentur, in *Concrete Science and Engineering—A Tribute to Arnon Bentur, Proceedings of the International RILEM Symposium 2004, Evanston, IL, USA, 22–24 March 2004*. Pumice Aggregates for Internal Water Curing (National Institute of Standards and Technology: Gaithersburg, MD, USA, 2004), pp 137–151. https://doi.org/10.1617/2912143586.013

X. Ma, J. Zhang, J. Liu, Review on superabsorbent polymer as internal curing agent of high performance cement-based material. China Ceram. Soc. **43**, 1099–1110 (2015). https://doi.org/10.14062/j.issn.0454-5648.2015.08.12

X. Ma, J. Liu, C. Shi, A review on the use of LWA as an internal curing agent of high performance cement-based materials. Constr. Build. Mater. **218**, 385–393 (2019). https://doi.org/10.1016/j.conbuildmat.2019.05.126

S.R.C. Madduru, P.R. Kumar, P.S.N.R. Giri, G.R. Kumar, Performance and microstructure characteristics of self curing self compacting concrete. Adv. Cem. Res. **30**(10), 451–468 (2018). https://doi.org/10.1680/jadcr.17.00154

V. Mechtcherine, H.W. Reinhardt, *Application of superabsorbent polymers (SAP) in concrete construction* (Springer, 2012)

References

V. Mechtcherine, M. Wyrzykowski, C. Schröfl, D. Snoeck, P. Lura, N. De Belie, S.I. Igarashi, Application of super absorbent polymers (SAP) in concrete construction. Update of RILEM state-of-the-art report. Mater. Struct. **54**(2), 1–20 (2021)

P.K. Mehta, P.J.M. Monteiro, Concrete-microstructure, properties and materials. BooksInBytes (2001)

H.D. Nguyen, H.Q. Le, in *IOP Conference Series: Materials Science and Engineering*. Water Movement in Internally Cured Concrete (2018), pp. 1–8. https://doi.org/10.1088/1757-899X/365/3/032029

S. Oh, Y.C. Choi, Superabsorbent polymers as internal curing agents in alkali activated slag mortars. Constr. Build. Mater. **159**, 1–8 (2018). https://doi.org/10.1016/j.conbuildmat.2017.10.121

F. Ridi, E. Fratini, P. Baglioni, Cement: A two-thousand-year-old nano-colloid. J. Colloid Interf. Sci. **357**, 255–264 (2011). https://doi.org/10.1016/j.jcis.2011.02.026

D. Sarbapalli, Y. Dhabalia, K. Sarkar, B. Bhattacharjee, Application of SAP and PEG as curing agents for ordinary cement- based systems: Impact on the early age properties of paste and mortar with water-to-cement ratio of 0.4 and above. Eur. J. Environ. Civil Eng. **21**, 1237–1252 (2016). https://doi.org/10.1080/19648189.2016.1160843

T. Sathanandham, R. Gobinath, M. Naveenprabhu, S. Gnanasundar, K. Vajravel, G. Sabariraja, R. Manoj, R. Jagathishprabu, Preliminary studies of self curing concrete with the addition of polyethylene glycol. Int. J. Eng. Res. Technol. **2**, 313–323 (2013)

D. Shen, X. Wang, D. Cheng, J. Zhang, G. Jiang, Effect of internal curing with super absorbent polymers on autogenous shrinkage of concrete at early age. Constr. Build. Mater. **106**, 512–522 (2016). https://doi.org/10.1016/j.conbuildmat.2015.12.115

D. Shen, C. Liu, J. Jiang, J. Kang, M. Li, Influence of super absorbent polymers on early-age behavior and tensile creep of internal curing high strength concrete. Constr. Build. Mater. **258**, 120068 (2020). https://doi.org/10.1016/j.conbuildmat.2020.120068

X. Sun, B. Zhang, Q. Dai, X. Yu, Investigation of internal curing effects on microstructure and permeability of interface transition zones in cement mortar with SEM imaging, transport simulation and hydration modeling techniques. Constr. Build. Mater. **76**, 366–379 (2015). https://doi.org/10.1016/j.conbuildmat.2014.12.014

T. Suwan, P. Wattanachai, Properties and internal curing of concrete containing recycled autoclaved acrated lightweight concrete as aggregate. Adv. Mater. Sci. Eng. 1–11 (2017). https://doi.org/10.1155/2017/2394641

M. Suzuki, M.S. Meddah, R. Sato, Use of porous ceramic waste aggregates for internal curing of high-performance concrete. Cem. Concr. Res. **39**, 373–381 (2009). https://doi.org/10.1016/j.cemconres.2009.01.007

P. Tang, M.V.A. Florea, H.J.H. Brouwers, Employing cold bonded pelletization to produce lightweight aggregates from incineration fine bottom ash. J. Clean. Prod. **165**, 1371–1384 (2017). https://doi.org/10.1016/j.jclepro.2017.07.234

P. Trtik, B. Münch, W.J. Weiss, A. Kaestner, I. Jerjen, L. Josic, E. Lehman, P.J.N.I. Lura, Release of internal curing water from lightweight aggregates in cement paste investigated by neutron and X-ray tomography. Nucl. Instrum. Methods Phys. Res. Sect. A **651**(1), 244–249 (2011). https://doi.org/10.1016/j.nima.2011.02.012

N. Tuan, G. Van Ye, B.K. Van, O. Copuroglu, Hydration and microstructure of ultra high performance concrete incorporating rice husk ash. Cem. Concr. Res. **41**(11), 1104–1111 (2011). https://doi.org/10.1016/j.cemconres.2011.06.009

F. Wang, Y. Zhou, B. Peng, Z. Liu, S. Hu, Autogenous shrinkage of concrete with super-absorbent polymer. ACI Mater. J. **106**, 123–127 (2009). https://doi.org/10.14359/56458

F. Wang, J. Yang, S. Hu, X. Li, H. Cheng, Influence of superabsorbent polymers on the surrounding cement paste. Cem. Concr. Res. **81**, 112–121 (2016). https://doi.org/10.1016/j.cemconres.2015.12.004

D. Wei, Y. Bin, L. Yibo, C. Liefang, Z. Yuncan, W. Yuan et al. Lightweight aggregates and its test methods. Part 2: test methods for lightweight aggregates (2010). https://www.chinesestandard.net/Index.aspx

H.S. Wong, N.R. Buenfeld, Euclidean distance mapping for computing microstructural gradients at interfaces in composite materials. Cem. Concr. Res. **36**(6), 1091–1097 (2006). https://doi.org/10.1016/j.cemconres.2005.10.003

J. Yang, F. Wang, X. He, Y. Su, Pore structure of affected zone around saturated and large superabsorbent polymers in cement paste. Cem. Concr. Compos. **97**, 54–67 (2019). https://doi.org/10.1016/j.cemconcomp.2018.12.020

S. Yehia, M. Alhamaydeh, S. Farrag, High-strength lightweight SCC matrix with partial normal-weight coarse-aggregate replacement: strength and durability evaluations. J. Mater. Civ. Eng. **26**(11), 04014086 (2014). https://doi.org/10.1061/(asce)mt.1943-5533.0000990

M.H. Zhang, O.E. Gjerv, Microstructure of the interfacial zone between lightweight aggregate and cement paste. Cem. Concr. Res. **20**, 610–618 (1990)

Y.Z. Zhuang, D.D. Zheng, Z. Ng, T. Ji, X.F. Chen, Effect of lightweight aggregate type on early-age autogenous shrinkage of concrete. Constr. Build. Mater. **120**, 373–381 (2016). https://doi.org/10.1016/j.conbuildmat.2016.05.105

S. Zhutovsky, K. Kovler, A. Bentur, Efficiency of lightweight aggregates for internal curing of high strength concrete to eliminate autogenous shrinkage. Mater. Struct./materiaux Et Constr. **34**(246), 97–101 (2002). https://doi.org/10.1007/bf02482108

S. Zhutovsky, K. Kovler, A. Bentur, Influence of cement paste matrix properties on the autogenous curing of high-performance concrete. Cem. Concr. Compos. **26**(5), 499–507 (2004). https://doi.org/10.1016/S0958-9465(03)00082-9

S. Zhutovsky, K. Kovler, Effect of internal curing on durability-related properties of high performance concrete. Cem. Concr. Compos. **42**, 20–26 (2012). https://doi.org/10.1016/j.cemconres.2011.07.012

D. Zou, K. Li, W. Li, H. Li, T. Cao, Effects of pore structure and water absorption on internal curing efficiency of porous aggregates. Constr. Build. Mater. **163**, 949–959 (2018). https://doi.org/10.1016/j.conbuildmat.2017.12.170

Chapter 5
Characteristics of Green Artificial Aggregates (GAA) as Self Curing Agent

5.1 Introduction

This chapter provides an examination of the physical and mechanical characteristics of GAA, focusing on its roles and performance as a self curing agent. This chapter starts by characterizing this green aggregate, analyzing their composition, size distribution, shape and surface texture. In parallel, granite, a conventional and widely used natural aggregate, renowned for its durability and strength was also examined for comparison. The chapter delves into a comparative study of these materials, highlighting their performance in terms of particle size distribution, density/specific gravity, water absorption, desorption and texture of aggregates. Additionally, their mechanical properties, such as Aggregate Crushing Value (ACV) is also explored. This chapter offers valuable insights into how integrating green alternatives and traditional granite to meet the demands of both modern engineering and environmental stewardship particularly to self cured concrete.

Artificial aggregates produced from several waste materials could be one of agents that could be used as self cured agent in concrete. GAA is a production from a combination of the wastes, for example, ceramic waste, bottom ash and fly ash. The influence of incorporating GAA as self curing agent to concrete properties and the mechanism under curing and without proper curing process is becoming the main interest of the authors. Within the knowledge of the authors, the application of artificial aggregate produced from combination of wastes to be used as self curing agent in concrete has not been reported elsewhere. Therefore, it is an intention of the authors to explore the utilization of GAA, one of lightweight aggregates (LWAs) as self curing agent in concrete. It also drives the authors to investigate the influence of incorporating GAA as self curing agent toward the densification internal microstructure for normal strength concrete by using Scanning Electron Microscopy (SEM), X-ray diffraction (XRD) and Mercury Intrusion Porosimetry (MIP) which will be elaborated in the following sections.

5.2 Physical Properties of GAA

The GAA was obtained from Active Pozzolan Technology Sdn. Bhd., Pasir Gudang, Johor, Malaysia. Prior to using GAA as a self cured agent, its physical properties are tested, namely particle size distribution by sieve analysis, density and specific gravity. The properties are conducted as according to ASTM C136 (2005) (sieve analysis) and ASTM C127 (2015) (density and specific gravity). Subsequently, aggregate absorption and desorption testing were conducted based on ASTM C127 (2015) and ASTM C1498 (2016), respectively, to confirm whether the GAA can store water and release the absorbed water in concrete mix. For comparison, all the tests were also carried out for granite. The following sub-sections elaborate each physical properties tested for GAA and granite.

5.2.1 Specific Gravity

Table 5.1 shows the physical and mechanical properties of GAA and granite and the standard used for testings. From the results, specific gravity of granite and GAA is found to be 2.6 and 1.9, respectively. The specific gravity for GAA is lower than specific gravity for the granite. Lower specific gravity of GAA by 28% from the granite is attributed to the presence of voids and air spaces within the GAA. The density of LWA is generally below 2000 kg/m^3 (Przychodzień and Katzer 2021; Demirboğa and Kan 2013; Durga et al. 2018; Chai et al. 2017). Therefore, GAA is classified as LWA.

Table 5.1 Physical, mechanical properties standard used for testings coarse aggregates

Properties	Granite	GAA	Standard for testing
Physical properties			
Specific gravity	2.65	1.92	ASTM C127 (2015)
Water absorption	0.32%	10.6%	ASTM C127 (2015)
Shape and surface texture	Angular and smooth	Spherical and rough	–
Mechanical properties			
Aggregate crushing value (ACV)	17.6%	24.35%	BS 812-110 (1990)
Aggregate crushing resistance (100-ACV)	82.4%	75.65%	

5.2.2 Water Absorption

Water absorption results for granite and GAA are tabulated in Table 5.1. Water absorption for granite and GAA are 0.32% and 10.6%, respectively. It shows that the water absorption for GAA is more than thirty (30) times higher than that of granite due to the porosity inside the GAA. In contrast, granite is almost no water absorption. Water absorption is an important property when considering the use of aggregates in concrete mixes, particularly in the context of self curing concrete (Ahmed et al. 2021; Grabiec et al. 2020).

The high-water absorption of GAA makes it a more suitable aggregate for self curing concrete. The porous nature of GAA allows it to absorb and store water during the mixing and casting concrete. Over time, as the concrete dries, the stored water can be gradually released from the GAA aggregates, helping to maintain the necessary moisture levels within the concrete for continued hydration and curing. Figure 5.1 depicts the colored pigment surrounding the GAA is absorbed water. The water absorption of GAA was in the range of 1–1.5 mm in depth after 24 h soaked in pigmented water.

5.3 Water Desorption

Since granite almost does not absorb water, the desorption test for granite is not performed. The desorption test result for sample of concrete containing GAA soaked in the 24 h pigmented water which conducted at 14 days is shown in Fig. 5.2. The distance of movement water from pre-wetted GAA to cement paste in the concrete specimen was in the average 0.77 mm. It shows that 60% absorbed water was released from the pre-wetted GAA. Figure 5.3 shows a desorption behavior of GAA exposed

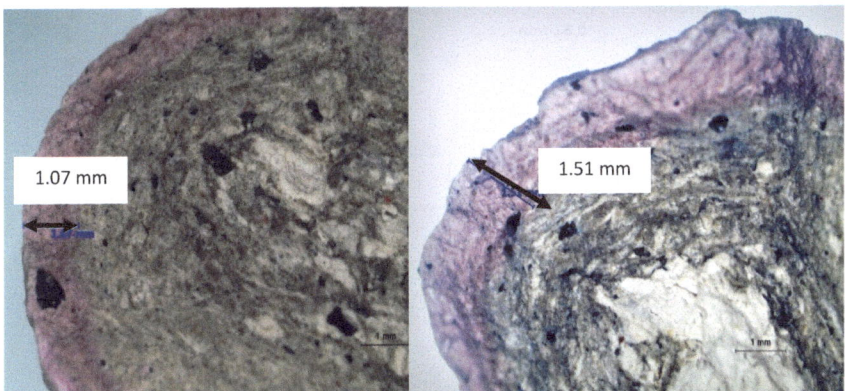

Fig. 5.1 Water absorption of GAA after 24 h soaked in pigmented water

at different RH. At 97% RH, the GAA released 52% of its absorbed water. The curve become steeper, almost plateau and steeper again, indicating a rapid rate in water release as RH reduces. The absorption for 24 h was selected as it is frequently used in practice by previous researchers to saturate LWAs for internal curing (Aghaee and Khayat 2023; Xu et al. 2021). According to ASTM C1761 (2017), LWA is acceptable for internal curing if absorbed water is released readily as the internal RH of the concrete reduces owing to self-desiccation. The water release rate of the aggregates are affected by RH and types of aggregates (Zou et al. 2018). Based on Figs. 5.2 and 5.3, it is indicated that GAA is a potential to be an internal curing agent in concrete due to its capability to release absorbed water from GAA. The water migration from GAA to cement paste is based on the law of fluid flow and the system's law of capillary attraction (Nguyen and Le 2018), whereby the radius of pores in cement paste is smaller than the pores in GAA. Capillary suction pressure generated by capillary pores in concrete is greater than the capillary suction pressure in GAA pores, causing of water movement from GAA to the concrete capillary pores for the curing process during the self-desiccation of concrete (Zhutovsky et al. 2011; Bandara et al. 2019). Therefore, the diagram of the movement of water from pre-wetted GAA to the cement paste is illustrated in Fig. 5.4. It can be concluded that GAA studied here are able to be self curing agent in concrete.

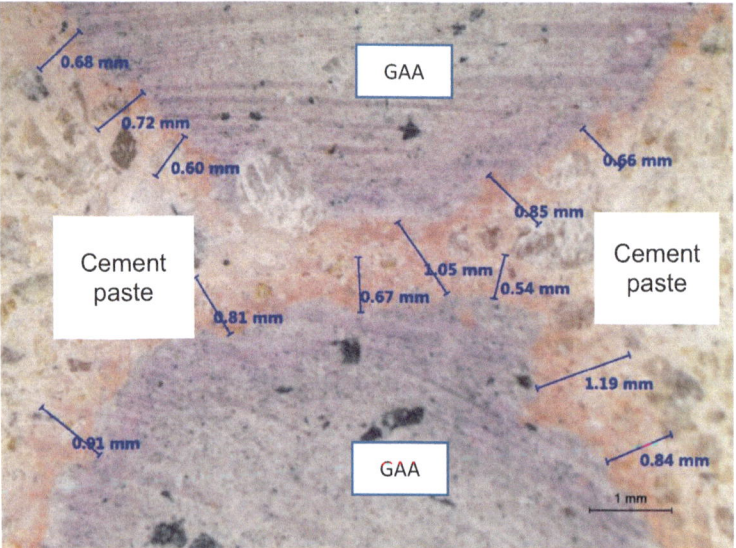

Fig. 5.2 The water movement in the concrete specimen from pre-wetted GAA to cement paste

5.4 Morphology and Microstructure of Aggregate

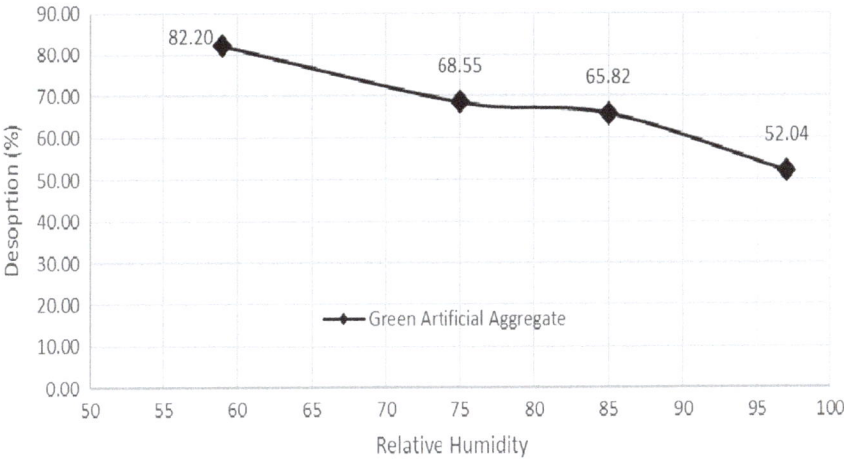

Fig. 5.3 Water desorption properties of pre-wetted GAA at different RHs

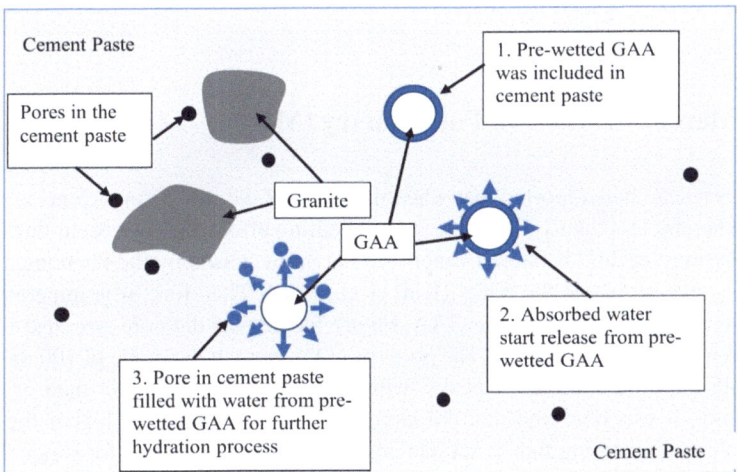

Fig. 5.4 The model illustrating the movement of water from pre-wetted GAA to cement paste

5.4 Morphology and Microstructure of Aggregate

Stereo Microscope was used to examine the morphological structure of GAA and granite. Figure 5.5 shows the stereo microscope image of GAA used in casting concrete. The GAA consists of spherical shape and rough surface texture compared to that granite which is irregular shape, and the texture is smooth. Spherical aggregates improve the workability of fresh concrete, while a rough surface of aggregate having larger contact area allows a better adherence to cement paste and increases

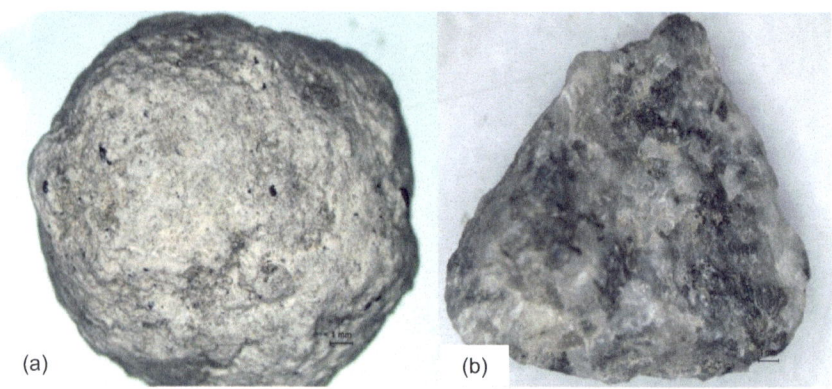

Fig. 5.5 Image of **a** GAA and **b** granite up to 4.5 × magnification

the splitting tensile strength of the concrete (Douglas and Garboczi 2007; Matias et al. 2013; Mehta and Monteiro 2014; Balapour et al. 2020) attributes to a smaller intrinsic viscosity than other shapes.

5.5 Mercury Intrusion Porosimetry (MIP)

Previous researchers reported that when the pores of the self curing agents are very small whereby less than 100 nm, the internal curing effect does not occur due to the difficulty in releasing the stored water into the paste caused by the phenomenon of capillary suction inside the pores (Kim et al. 2018). Therefore, it is imperative to examine the size of pores inside GAA. Figure 5.6 depicts the pore size distribution of GAA as determined by MIP. The pores of GAA were in the range of 100–600 nm size with multiple continuous peaks, with 52% pores being greater than or equal to 100 nm. It was observed that the larger the pore diameter, the higher the peak pore volume, indicating that GAA can act as an internal reservoir for water. These pores can store the amount of water and gradually release it into the cement paste to promote the continued hydration process. According to Gao et al. (2021), the pore size of granite showed multiple discontinuous peaks, indicating that the granite was a dense rock with poor pore connectivity as shown in Fig. 5.7.

5.6 Aggregate Crushing Value (ACV)

Aggregate crushing value (ACV) was carried out according to BS 812-110 (1990) to determine the strength of aggregate. Table 5.1 shows the result of ACV for GAA and granite. The ACV for GAA is 24.35%, which is higher than granite, recorded 17.6%

5.6 Aggregate Crushing Value (ACV)

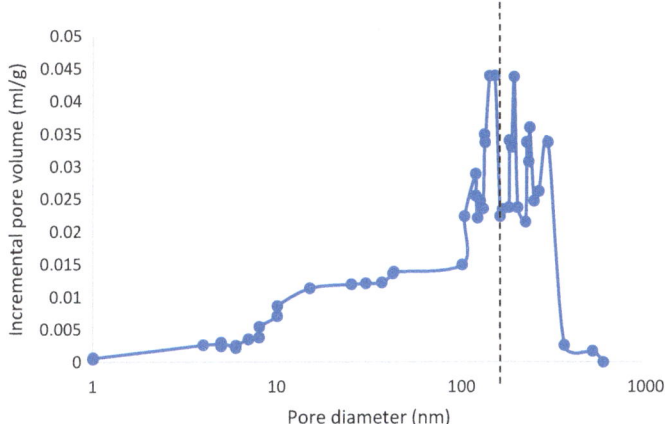

Fig. 5.6 Pore size distribution of GAA

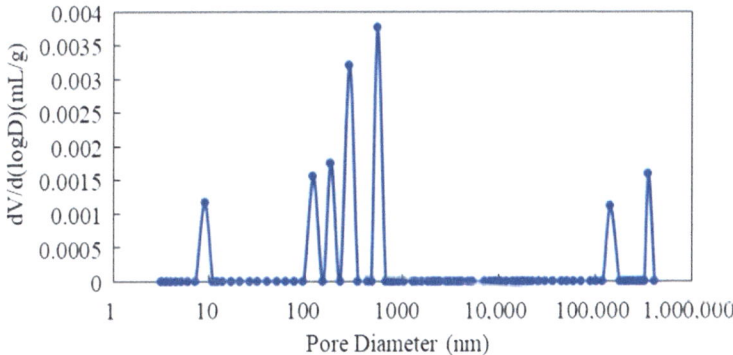

Fig. 5.7 Pore size distribution of granite (Gao et al. 2021)

ACV. This means that the GAA produces a higher percentage of debris particles than granite if the same compressive load is applied. This might be because of inherent porosity in GAA, which weakens the aggregate structure and makes it more prone to deformation and fragmentation under compressive loading conditions. Therefore, the aggregate crushing resistance of granite and GAA are 82.4% and 75.65%, respectively. It infers that granite possesses higher resistant toward crushing load. ACV is related on the strength of aggregates itself (Hossain et al. 2012; Iffat et al. 2017; Zhutovsky and Kovler 2017; Ma et al. 2019).

Previous researchers found that the strength of concrete reduced as the replacement of LWA increased which the latter associates with low strength (Grabiec et al. 2020; Rasheed and Abdulrasool 2020; Costa et al. 2012; Raoufi et al. 2011). According to BS EN 12620 (2002), aggregate with a crushing value below 25% is suited for heavy-duty floors, while a value beyond 30% is more appropriate for concrete wearing surfaces. Aggregate with a crushing value exceeding 45% can be utilized for various concrete purposes. Therefore, GAA is suitable for concrete that subject to wearing.

5.7 Initial Performance of GAA as Self Cured Agent in Concrete

The second stage of investigation is determination of initial performance for the concrete which the natural aggregates (granite) were replaced by 50% and 100% with GAA. Prior to mixing the aggregates in the concrete, the aggregates were soaked for 24 h to get the saturated-surface-dry (SSD) aggregate condition. Nevertheless, air-dry aggregates (AD) were also prepared to suit in-field applications and for comparison. The water cement ratio used was kept constant at 0.53 for all series of the concrete specimens. The concrete mix was designed according to Building Research Establishment (BRE) (Teychenné et al. 1997) method for grade 30. The aggregates used in the concrete mixes is divided into saturated-surface-dry (SSD) and air-dry (AD) and conditions. There are six (6) mixes of concrete and designated as follows:

(1) concrete containing 100% AD granite as control concrete (CTR100%-AD);
(2) concrete containing 100% SSD granite as control concrete (CTR100%-SSD);
(3) concrete containing 50% AD green artificial aggregate (GAA50%-AD);
(4) concrete containing 50% SSD green artificial aggregate (GAA50%-SSD);
(5) concrete containing 100% AD green artificial aggregate (GAA100%-AD); and
(6) concrete containing 100% SSD green artificial aggregate (100%GAA-SSD).

The workability of the fresh concrete mixes was performed based on BS 1881-102 (1983a). The specimen size 100 mm × 100 mm × 100 mm cubes were used to test compressive strength. The compressive strength concrete were carried out in accordance with BS 12390-3 (2019). Meanwhile, mode of failure was examined based on BS 1881-119 1983b). The concrete specimens were cured in water curing (WC). To simulate the improper curing conditions, a batch of the specimens were also air cured (AC) for 3, 7 and 28 days. The initial results for workability and compressive strength for control granite and concrete containing GAA are presented in the following sub-sections (Fig. 5.8).

5.7.1 Workability of Concrete with and Without GAA

The workability of the concrete mixes with constant water-cement ratio incorporating SSD and AD GAA in terms of slump is presented in Fig. 5.8. The data shows a decrease in slump as the proportion of AD GAA increases in the mixes. The slump of fresh concrete decreased by 26% when 50% of the granite was replaced with AD GAA and further reduced to 40% when 100% of the granite was replaced with

5.7 Initial Performance of GAA as Self Cured Agent in Concrete

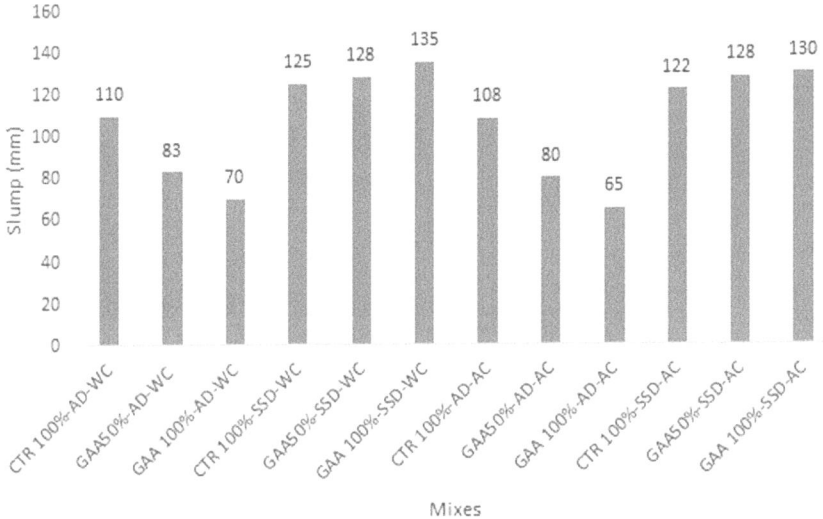

Fig. 5.8 Slump (in mm) of fresh concrete mixes containing 50 and 100% GAA which prior to that the aggregates were made SSD and AD conditions

GAA. The reduction in slump can be attributed to the water absorption capacity of AD GAA. Additional water is always needed during the mixing if LWA is used to produce fresh concrete with good workability (Tang et al. 2017). Conversely, it shows increase in slump as the proportion of SSD GAA increases in the mixes. The slump of fresh concrete increases by about 2–5% when 50% of the granite is replaced with SSD GAA and further rises by about 6–7% when 100% of the granite was replaced with GAA. This is because of the moisture that contains near to the surface of GAA for its SSD condition and their round shape improve the workability of the concrete mixes that made of such aggregate (Chinnu et al. 2021).

5.7.2 Compressive Strength of Concrete with and Without GAA

The compressive strength of the concrete specimens containing different percentages of AD GAA and SSD GAA, which were cured in water (WC) and air curing (AC) conditions for 3, 7 and 28 days, is shown in Fig. 5.9.

The control specimens made of 100% granite aggregates (CTR 100%) are compared against those made of GAA. It is revealed that the compressive strength decreases as the proportion of GAA increases. At the age of 28 days, the control specimens using 100% AD granite aggregates and cured in water (WC) demonstrate compressive strength that was 14% higher compared to the specimens made

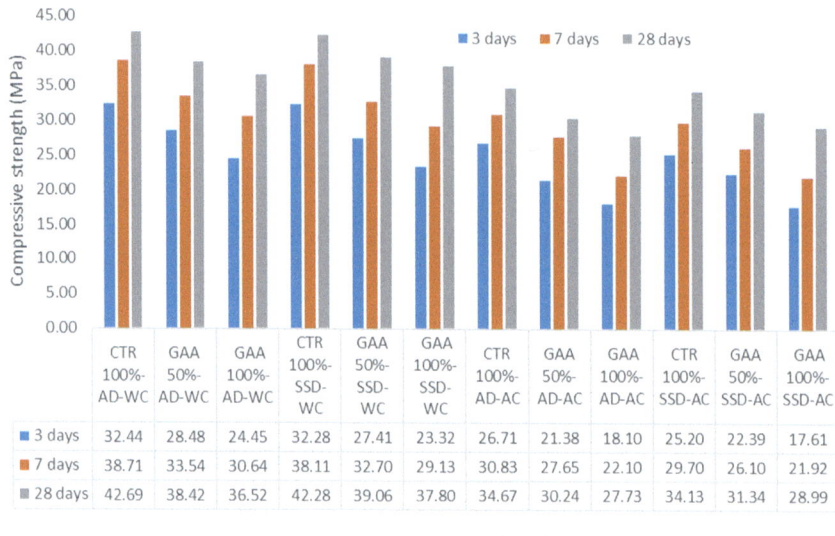

Fig. 5.9 Compressive strength of the concrete specimens containing SSD, AD GAA cured in air and water curing

of 100% AD GAA cured in water. The control specimens using 100% granite aggregates in a saturated surface dry (SSD) condition and cured in water showed an 11% higher compressive strength compared to those made of 100% SSD GAA and water-cured. It infers that the compressive strength of concrete contained SSD GAA boost its compressive strength closer to that using granite. Nevertheless, the compressive strength of 100% GAA-SSD-WC specimens from 7 to 28 days exhibited an increase of 22.94% in comparison to control concrete (CTR 100%-SSD-WC) with same condition. Meanwhile, CTR 100%-SSD-WC recorded increased by only 9.85% from 7 to 28 days. The CTR 100%-AD-WC specimens showed 9.31% and GAA 100%-AD-WC showed increased by 16.12% from 7 to 28 days. It infers that the water absorbed and retained in GAA acts as an internal reservoir, assisting the curing process. This effect is more pronounced for those that contained SSD GAA. It is evidenced that GAA with SSD condition prior to mixing does provide internal curing for promoting hydration from 7 to 28 days. The increment in compressive strength due to prolonged of curing times are for water cured and air cured are shown in Tables 5.2 and 5.3, respectively.

SSD aggregates enhance the compressive strength more effectively than those contained AD aggregates due to its consistent moisture content. Water curing significantly improves the compressive strength for both concrete containing granite and GAA aggregates, with a more pronounced effect seen in conventional granite aggregates. The GAA exhibits a notable increase in compressive strength over time, suggesting that while it may initially underperform compared to those concrete made of granite, it has a potential for substantial strength gain as it cures. By understanding

5.7 Initial Performance of GAA as Self Cured Agent in Concrete

Table 5.2 The percentage of increased concrete strength in 3 days, 7 days and 28 days cured in water curing (WC)

Types	Compressive strength (MPa)			Increase in strength from 3 to 7 days (%)	Increase in strength from 7 to 28 days (%)
	3 days	7 days	28 days		
CTR 100%-AD-WC	32.44	38.71	42.69	16.20	9.31
GAA 50%-AD-WC	28.48	33.54	38.42	15.09	12.71
GAA 100%-AD-WC	24.45	30.64	36.52	20.19	16.12
CTR 100%-SSD-WC	32.28	38.11	42.28	15.32	9.85
GAA 50%-SSD-WC	27.41	32.70	39.06	16.16	16.28
GAA 100%-SSD-WC	23.32	29.13	37.80	19.94	22.94

Table 5.3 The increase (%) in concrete strength in 3 days, 7 days and 28 days cured in air curing (AC)

Mix Designation	Compressive strength (MPa)			Increase in strength from 3 to 7 days (%)	Increase in strength from 7 to 28 days (%)
	3 days	7 days	28 days		
CTR 100%-AD-AC	26.71	30.83	34.67	13.38	11.07
GAA 50%-AD-AC	21.38	27.65	30.24	22.68	8.57
GAA 100%-AD-AC	18.10	22.10	27.73	18.10	20.30
CTR 100%-SSD-AC	25.20	29.70	34.13	15.14	12.98
GAA 50%-SSD-AC	22.39	26.10	31.34	14.21	16.73
GAA 100%-SSD-AC	17.61	21.92	28.99	19.67	24.37

these differences, engineers and construction professionals can make informed decisions on the choice of aggregates and curing methods to optimize the strength and durability of concrete structures.

Table 5.2 shows the increase (%) in compressive strength for all series of the specimens from 3 to 7 days and from 7 to 28 days, cured in air curing (AC). A similar trend of decreased compressive strength was observed at all ages as the proportion of granite was replaced by GAA. At the age of 28 days, the compressive strength of CTR 100%-AD-WC was higher 20% than those made of GAA in same condition. Meanwhile, the compressive strength for CTR 100%-SSD-WC specimens were higher 15% compared to those made of GAA with same condition. It is confirmed that concrete made of LWA always lower than that of control made of natural aggregates due to high volume of pores inside LWA.

It was observed that the compressive strength of GAA 100%-SSD-AC increased by a greater percentage of 24.37% from 7 to 28 days than that of the control concrete (CTR 100%-SSD-AC) which recorded 12.98% increase from 7 to 28 days. The CTR 100%-AD-AC recorded 11.07%, and GAA 100%-AD-AC recorded 20.30%, the highest increase from 7 to 28 days. The concrete specimens made of GAA still show higher increase from 7 to 28 days despite of its AD condition. The results may be attributed to the water absorbed and stored within GAA serving as an internal reservoir, gradually releasing water during the hydration process.

In sum, a self cured concrete could be developed based on GAA. This contributes to concrete sustainable material instead of chemical materials and material from natural resources. The use of GAA as a self curing agent in the concrete by completely replacing granite aggregate is considered a step toward reducing carbon emissions. Furthermore, it will minimize environmental impact by reducing the application of materials from natural resources and chemical materials. Then, it positively contributes to safeguarding and protecting the environment leading to sustainable development. Besides, the service life of the building structure would be enhanced by self curing technique compared to conventional non-cured concrete that eventually would reduce the cost of maintenance. This initial findings of the study are intended to encourage the implementation of new inventive approaches in production of concrete to be used in the building industry. This will help to promote better construction quality.

References

K. Aghaee, K.H. Khayat, Effect of internal curing and shrinkage-mitigating materials on microstructural characteristics of fiber-reinforced mortar. Constr. Build. Mater. **386**(February), 131527 (2023). https://doi.org/10.1016/j.conbuildmat.2023.131527

M. Ahmed, S. Alqadhi, S. Alsulamy, S. Islam, R.A. Khan, M. Danish, Development of self-cured sustainable concrete using local water-entrainment aggregates of vesicular basalt. Sustainability **13**(12) (2021). https://doi.org/10.3390/su13126756

ASTM C127, Standard test method for density (specific gravity) and absorption of coarse aggregate. ASTM Int. (2015)

ASTM C136, *Standard Test Method for Sieve Analysis of Fine and Coarse Aggregates* (2005)

ASTM C1761, *Standard Specification for Lightweight Aggregate for Internal Curing of Concrete* (2017). https://doi.org/10.1520/C1761

ASTM C1498, Standard test method for hygroscopic sorption isotherms of building materials. ASTM Int. **I** (2016)

M. Balapour, W. Zhao, E.J. Garboczi, N.Y. Oo, S. Spatari, Y.G. Hsuan, P. Billen, Y. Farnam, Potential use of lightweight aggregate (LWA) produced from bottom coal ash for internal curing of concrete systems. Cem. Concr. Compos. **105**, 103428 (2020). https://doi.org/10.1016/j.cemconcomp.2019.103428

M.M.H.W. Bandara, W.K. Mampearachchi, T. Anojan, Enhance the properties of concrete using pre-developed burnt clay chips as internally curing concrete aggregate. Case Stud. Constr. Mater. **11**, e00284 (2019). https://doi.org/10.1016/j.cscm.2019.e00284

British Method, BS 1881: Testing Concrete, *Part 102: "Method for Determination of Slump"* (1983a)

British Method, BS 1881: Testing Concrete, *Part 119: "Method for Determination of Compressive Strength Using Portions of Beams Broken in Flexure (Equivalent Cube Method)* (1983b).

British Method, BS 812-110, Methods of determination of aggregate crushing value (ACV) (Issue August) (1990)

British Standard, BS EN 12620:2002 +A1:2008. "Agregates for concrete". Incorporating corrigendum (2004)

BS EN 12390-3, 2019, TC: Testing hardened concrete—compressive strength of test specimens (2019)

References

L.J. Chai, H. Mahmud, M. Aslam, P. Shafigh, Effect of substitution of normal weight coarse aggregate with oil-palm-boiler clinker on properties of concrete. Sains Malaysiana **46**(4), 645–653 (2017). https://doi.org/10.17576/jsm-2017-4604-18

S.N. Chinnu, S.N. Minnu, A. Bahurudeen, R. Senthilkumar, Recycling of industrial and agricultural wastes as alternative coarse aggregates: a step towards cleaner production of concrete. Constr. Build. Mater. **287**(123056), 1–24 (2021). https://doi.org/10.1016/j.conbuildmat.2021.123056

H. Costa, E. Júlio, J. Lourenço, New approach for shrinkage prediction of high-strength lightweight aggregate concrete. Constr. Build. Mater. **35**, 84–91 (2012). https://doi.org/10.1016/j.conbuildmat.2012.02.052

R. Demirboğa, A. Kan, Design of specific gravity factor of artificial lightweight aggregate. Indian J. Eng. Mater. Sci. **20**(2), 139–144 (2013)

J.F. Douglas, E.J. Garboczi, *Intrinsic Viscosity and the Polarizability of Particles Having a Wide Range of Shapes*, vol. XC (2007). https://doi.org/10.1002/9780470141502.ch2.

J. Durga, C. Kumar, E. Arunakanthi, The use of light weight aggregates for precast concrete structural members. Int. J. Appl. Eng. Res. **13**(10), 7779–7787 (2018). http://www.ripublication.com

H. Gao, Y. Lan, N. Guo, Pore structural features of granite under different temperatures. Materials **14**(21), 6470 (2021). https://doi.org/10.3390/ma14216470

A.M. Grabiec, D. Zawal, W.A. Rasaq, The effect of curing conditions on selected properties of recycled aggregate concrete. Appl. Sci. **10**(13), 4441 (2020). https://doi.org/10.3390/app10134441

T. Hossain, A. Salam, M.A. Kader, Pervious concrete using brick chips as coarse aggregate: an experimental study. J. Civil Eng. **40**(2), 125–137 (2012). http://www.jce-ieb.org/pdfdown/4002003.pdf

S. Iffat, T. Manzur, M.A. Noor, Durability performance of internally cured concrete using locally available low cost LWA. KSCE J. Civ. Eng. **21**, 1256–1263 (2017). https://doi.org/10.1007/s12205-016-0793-x

J.H. Kim, S.W. Choi, K.M. Lee, Y.C. Choi, Influence of internal curing on the pore size distribution of high strength concrete. Constr. Build. Mater. **192**, 50–57 (2018). https://doi.org/10.1016/j.conbuildmat.2018.10.130

X. Ma, J. Liu, C. Shi, A review on the use of LWA as an internal curing agent of high performance cement-based materials. Constr. Build. Mater. **218**, 385–393 (2019). https://doi.org/10.1016/j.conbuildmat.2019.05.126

D. Matias, J. De Brito, A. Rosa, D. Pedro, Mechanical properties of concrete produced with recycled coarse aggregates - Influence of the use of superplasticizers. Constr. Build. Mater. **44**, 101–109 (2013). https://doi.org/10.1016/j.conbuildmat.2013.03.011

P.K. Mehta, P.J.M. Monteiro, *Concrete: Microstructure, Properties, and Materials*, 4th edn (2014)

H.D. Nguyen, H.Q. Le, Water movement in internally cured concrete. IOP Conf. Ser. Mater. Sci. Eng. 1–8 (2018). https://doi.org/10.1088/1757-899X/365/3/032029

P. Przychodzień, J. Katzer, Properties of structural lightweight aggregate concrete based on sintered fly ash and modified with exfoliated vermiculite. Materials **14**(20) (2021). https://doi.org/10.3390/ma14205922

K. Raoufi, J. Schlitter, D. Bentz, J. Weiss, Parametric assessment of stress development and cracking in internally cured restrained mortars experiencing autogenous deformations and thermal loading. Adv. Civil Eng. **2011**(1), 870128 (2011). https://doi.org/10.1155/2011/870128

L.S. Rasheed, A.T. Abdulrasool, Recyclable wastes as internal curing materials to improve high-performance concrete's sustainability, and durability: An overview. IOP Conf. Ser. Mater. Sci. Eng. **928**(2), 022071 (2020). https://doi.org/10.1088/1757-899X/928/2/022071

P. Tang, M.V.A. Florea, H.J.H. Brouwers, Employing cold bonded pelletization to produce lightweight aggregates from incineration fine bottom ash. J. Clean. Prod. **165**, 1371–1384 (2017). https://doi.org/10.1016/j.jclepro.2017.07.234

D.C. Teychenné, R.E. Franklin, H.C. Erntroy, D.W. Hobbs, B.K. Marsh, *Design of Normal Concrete Mixes* (IHS Building Research Establishment (BRE), BRE Press, UK, 1997)

F. Xu, X. Lin, A. Zhou, Performance of internal curing materials in high-performance concrete: a review. Constr. Build. Mater. **311**(December 2020), 125250 (2021). https://doi.org/10.1016/j.conbuildmat.2021.125250

S. Zhutovsky, K. Kovler, Influence of water to cement ratio on the efficiency of internal curing of high-performance concrete. Constr. Build. Mater. **144**, 311–316 (2017). https://doi.org/10.1016/j.conbuildmat.2017.03.203

S. Zhutovsky, K. Kovler, A. Bentur, Revisiting the protected paste volume concept for internal curing of high-strength concretes. Cem. Concr. Res. **41**(9), 981–986 (2011). https://doi.org/10.1016/j.cemconres.2011.05.007

D. Zou, K. Li, W. Li, H. Li, T. Cao, Effects of pore structure and water absorption on internal curing efficiency of porous aggregates. Constr. Build. Mater. **163**, 949–959 (2018). https://doi.org/10.1016/j.conbuildmat.2017.12.170

Index

A
Absolute, 55
Absorbed, 18, 22, 28–30, 33, 34, 39, 54, 57, 72–74, 80, 82
Absorption, 21, 22, 26, 27, 29, 32, 33, 35, 37, 53, 54, 56, 57, 63, 71–73, 79
Admixture, 2, 14, 18, 25, 33
Afro-asian, 21
Agents, 2, 4, 5, 18, 21, 22, 25, 26, 30, 33, 37, 53, 63, 71, 76
Aggregate crushing resistance, 72, 77
Aggregate crushing value, 71, 72, 76
Aggregates, 1, 2, 4, 5, 21, 22, 25–27, 32–36, 38, 39, 42, 53–56, 59, 60, 63, 71, 73–75, 77–81
Air-dry, 54, 58, 78
Alite, 12
Aluminate, 12, 16
Aluminate monosulfate, 16
Amorphous, 15
Angular, 55, 72
Angularity, 55
Artificial aggregates, 21, 71
Attraction, 26, 30, 74
Autogenous, 19, 28–31, 55–59

B
Belite, 12
Binders, 60
Binding, 15, 42, 55
Biochar, 27
Bond, 30, 42, 55, 59, 62
Bottom ash, 26, 27, 37, 41, 71
By-products, 21, 26

C
Calcium hydroxide, 12–15, 17, 59, 60
Capillary, 3, 17, 19, 26, 29, 31, 55, 56, 58, 74, 76
Carbon, 21, 22
Cement, 1–5, 11–19, 22, 26–29, 31, 33, 35, 37, 39, 40, 42, 53–57, 59–63, 73–76, 78
Cementitious, 2, 12, 18, 27, 29, 38
Cenosphere, 27
Ceramic, 5, 31, 71
Ceramic waste, 5, 32
C-H, 12, 14–17, 35, 63
Change, 11, 14, 28, 31, 55, 60
Chemical admixture, 2, 18, 25
Chemical composition, 18, 63
Chemical properties, 53, 64
Chemical reaction, 2, 11, 12, 14, 17, 19, 22, 29
Chemical shrinkage, 19, 55, 56, 58
Chemo thermal, 61
Clay, 5, 26, 27, 34–38
Compact, 55
Compounds, 2, 3, 11, 12, 62
Compressive, 2, 25, 33–40, 55, 77–81
Compressive strength, 35–38, 40, 78–81
Concrete, 1–5, 11–13, 15, 17–22, 25–41, 53–64, 71–75, 77–82
Condensation, 19, 30
Connectivity, 76
Contact, 20, 75
Continuous, 1–3, 19, 53, 76
Conventional, 1, 2, 4, 5, 11, 21, 22, 33–35, 38, 42, 53, 56, 59–61, 64, 71, 80, 82

Conventional aggregate, 33, 35, 38, 42, 53, 56, 59–61
Conventional curing, 2, 6
Corrosion, 16
Cracking, 3, 5, 17, 53, 55–57
Critical pore, 54
Cross-links, 29
Crushed, 31, 55
Crystallinity, 63
Crystals, 12, 16
C_2S, 15, 18
C_3S, 16, 18
C-S-H, 2, 12, 14–17, 35, 60, 62–64
Curing, 1–6, 11, 18–22, 25–30, 33, 34, 37–42, 53, 54, 56–64, 71, 73, 74, 78–81

D

Deformation, 53, 55, 64, 77
Degradation, 17
Degree, 35, 57, 61, 62
Deleterious, 2
Dense, 2, 17, 21, 62, 76
Density, 17, 25–27, 31–35, 38, 53, 72
Desorption, 53, 54, 57, 71–73, 75
Dicalcium silicate, 12, 14, 15
Diffusion, 19, 31
Dissolution, 11, 14
Drying shrinkage, 19, 53, 55–58
Durability, 1, 2, 5, 11, 15–18, 21, 22, 25, 29, 41, 53, 59, 61, 64, 71, 81

E

Early-age, 3, 17, 38
Effect, 3–5, 11, 21, 25, 30–33, 37, 38, 40, 41, 53–60, 62, 76, 80
Efficiency, 4, 11, 21, 28, 42, 54, 56, 62
Elongated, 55
Ethylene oxide, 30
Ettringite, 12, 16, 59, 60, 63
Evaporate, 57
Evaporation, 2–4, 18, 30, 37, 58
Expanded clay, 26, 27, 34–38
Expanded shale, 5, 27, 35–37, 58

F

Ferrite, 13
Fine, 1, 27, 32, 36, 56–59, 63
Flexural strength, 25, 33, 38, 40–42, 55
Flexure, 55
Flora, 21

Fly ash, 5, 26, 36, 71
Formwork, 2, 3
Freshwater, 21, 29

G

Gel, 15, 16, 29, 35, 62
Gradation, 55
Gradient, 19, 26, 29, 31, 58
Granite, 4, 33, 34, 39, 71–73, 75–78
Green aggregate, 22, 71
Green artificial aggregate, 5, 78
Gypsum ($CaSO_4 \cdot 2H_2O$), 12

H

Hardened, 2, 11–13, 16–18, 22, 34, 54, 60
Heat, 3, 15, 62
Heavy-duty, 77
High-performance, 4, 18, 34, 55, 57, 58, 62
High-Performance Concrete (HPC), 4, 34, 56, 57
Homogeneousness, 60
Humidity, 19, 26, 29
Hydration, 2–5, 11–19, 22, 25–28, 30, 32, 35, 37, 38, 41, 42, 55–58, 60–64
Hydraulic model, 61
Hydrogarnet, 13
Hydrogels, 29, 38
Hydrophilic, 25, 30, 62
Hydrophobic, 62

I

Impact, 1, 11, 22, 59
Induction, 14
Infrared curing, 3
Infrastructure, 2, 64
Integrity, 5, 55
Intensities, 63
Interfacial Transition Zone (ITZ), 53, 59
Interlinking, 63
Internal, 1, 3–5, 18–22, 25, 26, 28, 30, 37, 40–42, 53, 54, 56–58, 60–63, 71, 74, 76, 80, 82
Internal curing, 4
Intrinsic, 38, 76
Irregular, 75

K

Kinetics, 5

Index

L
Late shrinkage, 57
Lifecycle, 22
Lightweight, 4, 25, 26, 33–37, 42, 53
Lightweight Aggregate (LWA), 4, 5, 26, 27, 29, 33–38, 40, 42, 53, 56–60, 62, 63, 71, 72
Low-water absorption, 57

M
Mass, 27, 29, 55
Matrix, 2, 5, 17, 19, 31, 42, 57, 59–62
Mechanical, 2, 4, 11, 16, 25, 41, 42, 59, 60, 71, 72
Mechanical properties, 71
Membrane, 1, 3
Membrane curing, 1, 3
Mercury Intrusion Porosimetry (MIP), 5, 71, 76
Mesopores, 28
Mesoporous, 27
Mesoscale, 61
Microstructure, 2, 5, 11, 15, 17, 28, 53, 59–64, 71
Mitigate, 2, 4, 18, 29, 53, 55
Moisture, 1–5, 18, 21, 25, 31, 33, 53, 58, 59, 73, 79, 80
Monosulfate hydrate, 12
Monosulfoaluminate, 16
Morphology, 59, 75
Mortar, 21, 25, 33, 41, 42, 59–61
Movement, 19, 31, 73–75

N
Nano-structured, 16
Natural aggregate, 4, 21, 71, 78, 81
Non-cured, 21, 82
Normal-weight aggregates, 27, 39

O
OPC, 12
Osmotic, 29
Oven-dry, 54, 56, 57

P
Particle size distribution, 71, 72
Paste, 1, 2, 12–17, 19, 26–29, 33–35, 42, 53, 54, 56, 57, 59–61, 63, 73–76
Peaks, 63, 76
Permeability, 17, 60

pH, 15, 16, 29
Physical, 4, 12, 25, 39, 53, 58, 60, 61, 64, 71, 72
Physical properties, 39, 53, 64, 72
Pigmented, 73
Plastic, 2, 56–58
Plastic stage, 56–58
Polymers, 5, 25, 30
Pore, 2, 15–17, 26–29, 31, 35, 54, 56–62, 74, 76, 77, 81
Porosity, 2, 3, 17, 21, 26, 27, 31, 42, 53–55, 57, 59, 60, 62, 73, 77
Porous, 2, 17, 18, 25, 27, 28, 31, 33–35, 37, 41, 42, 54, 56, 60, 61, 73
Portlandite, 16, 63
Powders, 25, 27
Pre-induction, 14
Pre-saturated, 5, 19, 33
Pre-soaked, 41
Pressure, 27, 29–31, 74
Pre-wetted, 26, 34, 35, 54, 56, 57, 73–75
Pumice, 5, 27, 37, 42, 56, 63

R
Radius, 26, 74
Raoult's law, 30
Rate, 2, 14, 15, 17, 18, 21, 30, 32, 33, 56, 57, 63, 74
Relative humidity, 2, 26
Released, 18, 22, 27, 31, 54, 56, 73, 74
Reservoir, 4, 5, 18, 19, 26, 28, 29, 31, 37, 76, 80, 82
Resistant, 77
Resources, 21, 82
Retention, 4, 18, 22, 54
Rice husk ash, 27
Rock, 76
Rough, 26, 55, 72, 75
Rougher, 55
Roughness, 59
Round, 79

S
Sand, 4
Saturated-Surface-Dry (SSD), 58, 78
Saturation, 28, 54
Scanning Electron Microscopy (SEM), 5, 28, 60, 61, 71
Self curing, 1, 4–6, 11, 18, 19, 21, 22, 25–27, 29–31, 35, 36, 38, 41, 53, 54, 56, 57, 60, 71, 73, 74, 76, 82

Self-desiccation, 4, 19, 28, 30, 31, 56, 59, 74
Shale, 5, 26, 27, 35–37, 41, 58
Shape, 3, 26, 33, 53, 55, 59, 60, 71, 72, 75, 79
Shrinkage, 2, 3, 5, 19, 28, 29, 31, 37, 53, 55–58
Shrinkage-induced cracking, 53
Sieve analysis, 72
Silica fume, 26
Slate, 5, 26
Slump, 33, 34, 78, 79
Smooth, 55, 72, 75
Soaked, 33, 39, 41, 54, 63, 73, 78
Specific gravity, 34, 35, 37, 71, 72
Spectrum, 63
Spherical, 26, 34, 72, 75
Splitting, 40, 76
Standard gradation, 55
Steady-state, 15
Steam curing, 1, 3
Stereo, 75
Strength, 2, 5, 11, 12, 14–19, 22, 25, 32–42, 53–57, 59–64, 71, 76, 78–81
Structure, 2, 11, 12, 15–17, 27, 29, 33, 54, 56, 57, 75, 77, 81, 82
Suction, 19, 26, 74, 76
Sulfate, 12, 16
Sulfoaluminate, 12
Superabsorbent, 5, 25
Superabsorbent polymers (SAPs), 5, 29
Superfine, 25, 27
Superplasticisers, 18
Supply, 1, 4, 5, 16, 27, 53
Surfaces, 3, 78
Sustainability, 1, 4, 21, 22
Sustainable, 1, 6, 21, 22, 64, 82

T
Tensile strength, 38, 40, 41, 76
Tetracalcium Aluminoferrite (C_4AF), 12

Texture, 53, 55, 59, 71, 72, 75
Thermal, 17, 61
Threshold, 55
Transport properties, 61
Tricalcium Aluminate (C_3A), 12, 16
Tricalcium Silicate (C_3S), 12, 14

U
Ultra High Performance Concrete, 28, 39
Uncrushed, 55

V
Van der Waals, 12
Vapor, 27
Viscosity, 76
Voids, 12, 29, 37, 38, 55, 72
Volume, 12, 13, 28, 29, 34, 54, 55, 60, 76, 81

W
Waste, 5, 21, 22, 26, 27, 31, 32, 71
Water absorption, 26, 27, 32, 33, 35, 37, 53, 54, 56, 57, 63, 71–73, 79
Water binder, 19, 56
Water binder ratio (w/b), 19
Water-cementitious, 18
Water-cement ratio, 17, 18
Water cured, 38, 80
Water-filled, 4, 18
Water-retaining agents, 53
Water retention, 4, 18, 22, 54
Wearing, 78
Well-hydrated, 17, 18
Workability, 17, 18, 25, 30, 33, 34, 75

X
X-Ray Diffraction (XRD), 5, 62, 63

The manufacturer's authorised representative in the EU is Springer Nature Customer Service Centre GmbH, Europaplatz 3, 69115 Heidelberg, Germany. If you have any concerns regarding our products, please contact ProductSafety@springernature.com

Printed and bound by CPI Group (UK) Ltd, Croydon, CR0 4YY

26/03/2026

02078983-0003